An Introduction to
General Systems Thinking

Other Books by Gerald M. Weinberg

1961   *Computer Programming Fundamentals* (WITH H. D. LEEDS) (McGRAW-HILL)

1966   *PL/I Programming Primer* (McGRAW-HILL)

1970   *Computer Programming Fundamentals, Based on the IBM System/360* (WITH H. D. LEEDS) (McGRAW-HILL)

1970   *PL/I Programming—A Manual of Style* (McGRAW-HILL)

1971   *The Psychology of Computer Programming* (VAN NOSTRAND REINHOLD)

1973   *Structured Programming in PL/C* (WITH N. F. YASUKAWA AND - R. MARCUS) (WILEY)

1973   *Teacher's Guide to Structured Programming in PL/C* (WITH - N. F. YASUKAW AND R. MARCUS) (WILEY)

An Introduction to General Systems Thinking

Gerald M. Weinberg

A WILEY-INTERSCIENCE PUBLICATION

JOHN WILEY & SONS
New York • Chichester • Brisbane • Toronto • Singapore

Boys and young men acquire readily the moral sentiments of their social milieu, whatever these sentiments may be. The boy who has been taught at home that it is wicked to swear, easily loses this belief when he finds that the schoolfellows whom he most admires are addicted to blasphemy.

Bertrand Russell

To Ross Ashby, Kenneth Boulding, and Anatol Rapoport, who got me addicted to blasphemy.

Library of Congress Cataloging in Publication Data

Weinberg, Gerald M

An introduction to general systems thinking.

(Wiley series on systems engineering and analysis)
"A Wiley-Interscience publication."
Includes bibliographical references and indexes.
1. System theory.   I. Title.

Q295.W44   1975   003   74-26689
ISBN 0-471-92563-2

Printed and bound in the United States of America by Braun-Brumfield, Inc.

20 19 18 17 16

SYSTEMS ENGINEERING AND ANALYSIS SERIES

In a society which is producing more people, more materials, more things, and more information than ever before, systems engineering is indispensable in meeting the challenge of complexity. This series of books is an attempt to bring together in a complementary as well as unified fashion the many specialties of the subject, such as modeling and simulation, computing, control, probability and statistics, optimization, reliability, and economics, and to emphasize the inter-relationship between them.

The aim is to make the series as comprehensive as possible without dwelling on the myriad details of each specialty and at the same time to provide a broad basic framework on which to build these details. The design of these books will be fundamental in nature to meet the needs of students and engineers and to insure they remain of lasting interest and importance.

Preface

I found everything perfectly clear, and I really understood absolutely nothing. To understand is to change, to go beyond oneself. This reading did not change me.*†

This book is based on a course that over the years has changed the thinking of many people. In case you think that you are not the type of person to be changed by reading a book, let me quote for you some typical comments received in course evaluations.

An electrical engineer said of the course, "It made the many isolated subjects I had studied in college come together into a meaningful whole—and it also related them to my five years of on-the-job experience."

An archaeologist said, "I don't think I ever understood before the role of theory in my work, and just how powerful theory can be if you don't let it master you. When I dig now, I have always in my mind a perception of the site as a whole, and as a part of a larger whole, a living culture."

A composer said, "I probably couldn't demonstrate this to you exactly, but my recent compositions have been altered, definitely altered, and for the better, as a result of taking this course."

A computer systems analyst said, "I should have taken this course a dozen years ago. In three months I have learned more about what systems are than I knew previously. A problem that came up in my job and that would have caused me much grief was just erased with no effort because I was able to apply the Principle of Indifference. In another case, something that a few months ago would have slipped by unnoticed and gotten us into a lot of trouble was caught just because I

---

* It would be out of keeping with the informal tone of this book to clutter the pages with footnotes and references. We shall therefore confine all the other notes to the end of the book.

† Jean-Paul Sartre, *Search for a Method.* Translated by Hazel E. Barnes. New York: Vintage, 1968, pp. 17, 18. Sartre is referring to *Capital* and *German Ideology,* by Karl Marx.

almost unconsciously played some observer games with it. Under one of the new points of view, the problem was obvious. So was the solution."

But a computer programmer said, "I didn't learn anything in this course. It was a bunch of platitudes, no more than ordinary common sense. It was fun, but otherwise a waste of time."

You can't teach all of the people all of the time. We start with some promise of success and some warning that success is not guaranteed. To make things worse, books about thinking are a pox on the market—those who can't think write books about thinking. So beyond the several hundred testimonials running about 9 to 2 in favor of a significant change in thinking, what promise is there that this book can change your thinking and your understanding of the thoughts of others? Scholars learn to think in at least two distinct ways. One method begins with the mastery of the details of a discipline and then proceeds to transcend them. We speak of this transcendence in such approving terms as, "thinks physically," "knows anthropological theory," or "has mathematical maturity." What have we done in attaining this disciplinary maturity? For one thing, we have learned how to "approach" a problem—that is, what should be our first few thoughts.

This disciplinary method of teaching works well. First—obviously— it builds on the foundations of wisdom left by others and conserves the effort of retracing their steps. Second—and in our fragmented society, not so obvious—the disciplinarian confines himself to a rather small range of "problems," a range in which he is fairly confident of his ability to get results. A successful disciplinarian knows what problems to avoid.

But what of problems that refuse to be avoided? What of the depletion of our natural resources by an ever-increasing population in an ever-more-wasteful economy? What of expanding technology, usually the obedient servant but occasionally the terrible master? What of grisly wars and impoverished peace? What of death, and what of me, dying?

Such problems fall outside any discipline. Many lesser problems too come supplied with no familiar label. This book attempts to teach an approach to thinking when the labels are missing, or misleading. This approach *precedes* the disciplinary studies—and sometimes bypasses them, or integrates them. We call this way of thinking and teaching the general systems approach.

The general systems approach is not *my* invention. Many people have made original contributions to the general systems approach, but

I am not among them. Why, then, do I write this book? Only because, through a dozen years of attempting to teach general systems thinking I have found that none of the "introductory" books make it accessible to a truly general audience.

My role, consequently, is to integrate a mass of material into an introductory form. I have tried to gather insights both from general systems theorists and from disciplinarians, to arrange them in a consistent and helpful order, and to translate them into a simpler and more general language so that they become common property.

There is, then, a double meaning to the word "general" in the title: *the most generally applicable insights made available to the most general audience possible.*

By elevating particular disciplinary insights to a general framework and language, we make some ideas of each discipline available for the use of all. If these ideas have been well chosen to have general application, then this approach should yield for the disciplinarian a certain economy of thought—he need not retrace steps taken in other disciplines. This book, then, is not for "systems specialists," but for systems generalists.

Who are those "generalists"? Certainly they include—and have included in my courses over the years—almost anybody who uses his or her brain to make a better living, or to make living better. I have had managers and other organizational leaders, social and biological scientists, computer systems designers, many engineers, and a whole host of college undergraduates in all fields. I have had anthropologists and actors, businessmen and biologists, cartographers and cab drivers, designers and dilettantes, electrical engineers and Egyptologists, French majors and farmers—we need not continue the exercise.

Few of these people had mathematical training much beyond high-school algebra, and some not even that. The treatment of mathematical subjects in the book is geared to this level because it is the level on which most people—most educated people—happen to find themselves. A control systems engineer who reviewed this book felt a danger that, should his students read it "they would not want to study their calculus and differential equations."

But read what a chemistry student said: "The follow-on for this course for me is a course in differential equations. I always dreaded the thought after finishing calculus, and since it wasn't *required,* I just kept putting it off. But I knew I needed it, vaguely, and now I know why I need it *precisely.* More than that, I've lost my fear—they can't touch me now that I know what it's about." Or a sophomore biologist: "I haven't taken any math since high-school algebra. That's really stupid for a biologist, but until this general systems course I never knew that. I'll start calculus next semester, if they'll let me."

Can these claims be true? Leaf through the book, and you will find a variety of graphs, diagrams, symbols, and even equations. But they are not there to mystify. Just *because* ordinary people are so often alienated from science and technology by such devices, a book on general systems thinking must be designed to lift the veil off their mysteries.

The appropriate mathematical symbolism will first be justified, then explained, as needed. Contrary to popular belief, scientists use mathematics to make things clearer, not more obscure. I intend to use math only that way, so, if you find the symbolism unclear, try once more. If it is still unclear, give up, blame it on me, and proceed. You won't miss too much.

Not all sciences confound with mathematical symbols. Ordinary words do quite nicely—especially if you don't really know what you are talking about. My computer experiences have made me aware that people often have but a foggy idea of what they are saying. Through translating thoughts into computer programs, I have learned many fog-clearing techniques. These techniques would have been impossible without the knowledge gained from computing, which is why so few of them are understood by older scientists—and systems theorists. This book will not teach you to program computers, but it will teach you to think the way a computer programmer should.

And speaking of fog, let us leave no illusions about the clarity of my own thoughts. Over the years of writing, entire sections of this book have been scrapped as the mist has been dispelled. Moreover, I am not afraid to employ slight inaccuracies to make the lessons more forceful and therefore more memorable. In other words, I choose vigor over rigor.

So do not take this book too seriously. It is not a bible, nor a proof, nor even a cohesive argument. It is, indeed, *my* first few thoughts, a collection of hints, nudges, pushes, and sometimes shoves, which aim to assist *your* first few thoughts on any "systems" problem. As another of my students said, "I feel that this course has made me twice as good a (computer) systems designer, but I *know* it has made me ten times as good a thinker." I hope it will do as much for you. It may do more.

Gerald M. Weinberg

June 1974

How To Use This Book

In manuscript form, this book has been used in several ways, but particularly for individual or class use. Although the reader will undoubtedly discover his or her own ways of using it, some notes on how the author has seen it used and planned its use might be in order.

For individual use, the best approach is probably just to read it straight through, ignoring all the bibliographic material. The Questions for Further Research at the end of each chapter should probably be read as part of the text, to give an impression of the scope of problems to which the chapter materials might apply. Should some problem or quotation strike you as particularly intriguing, make note of it and then use the references to take it up later. Since the book is intended to introduce you to new ways of thinking, many quotations and references have been given—not to lend a patina of scholarship, but to give you numerous pointers toward other paths to learning.

Not all of the references represent *good* examples, so further assistance is given through the medium of the Recommended and Suggested Readings at the end of each chapter. The fundamental difference between "recommended" and "suggested" is that over the years I have found it imprudent to "recommend" that someone read an entire book. Either they don't do it, or they do and have a rather different perception than mine of its worth. In the latter case, I have made an enemy; in the former, I have made someone wish to avoid me. But do read some of the suggested books anyway.

For classroom use, there is a great variety of options. For a typical university curriculum, the seven chapters may be assigned one roughly every other week, with intermediate weeks used for the recommended readings. This is the scheme we use when we are dealing with a "mixed" audience—that is, with students from a variety of disciplines all in one class. When the class is more homogeneous, more specialized readings may be substituted. This approach has been used, to our knowledge, at least in management science, computer science, and behavioral science.

The text itself is suitable for any "level," from sophomores on up, with the adjustment being made by assignment of differing amounts of supplementary reading and questions for further research. The research questions themselves are usually suitable for either a short essay or a term paper. In higher-level courses, we have students prepare one or more of these questions for class presentation. For those students without mathematical background, the notational exercises are highly recommended.

The very flexibility of this book and generality of its material make it difficult to set in a university curriculum. "In what department does it belong, really?" "At what level student is it aimed?" These questions, so frequently asked of me, might be symptomatic of the excessive categorization of our society—breaking down knowledge into disciplinary fiefdoms and people into age-graded human waves passing through an education factory. But those who ask are often sincere in their attempt to cut through the present structure and obtain something better, and we should try to give them a helpful answer.

With regard to "where," a course or sequence in the general systems approach might be found in any department where there is a willing instructor and a cooperative chairman. In some places, cross listing of a course is the traditional way of handling such hot potatoes; in others, provision already exists for all-university, or at least all-division courses. Quite often, the Philosophy Department would be an appropriate place, except that our ex-resident philosopher, Virg Dykstra, always taught us that there shouldn't be a Philosophy Department—just a philosopher in every department. So perhaps there ought to be a general systems course in every department, taught by its philosopher. Alternatively, the book can certainly be supplementary reading in a variety of courses.

With regard to "who" or "when," I can be more specific if my own personal prejudices may be allowed to creep out. I have taught this material to sophomores, juniors, seniors, beginning and advanced graduate students, as well as to those long out of formal education. For some reason, the most exciting times were with seniors or those long out of school. The seniors seem to be looking for a way of integrating a befuddling mass of four years' worth of factual material into something they can actually *use*. Although at first glance the idea of this material being *useful* might seem hilarious, more than a few students have returned or written to tell me that this was the most useful course they took in four years of college. I hope that speaks well for the course and not badly for the college.

Perhaps this practicality is what makes the course take so well with working people, whose consistent reaction is to bring tales to class of how they applied, or should have applied, some general systems law or other to their daily labors. On the other hand, beginning graduate students seem too often obsessed with achieving the maximum specialization possible in the minimum time, while sophomores just want a few specifics on which to hang their generalities. But, naturally, people don't fall so neatly into these class categories, and I'd hate to think of the learning I would have missed by excluding certain graduate students and sophomores from my classes.

Acknowledgments

This book is the work of many people, work that merely happens to be assembled by one. First are the students who found themselves used as guinea pigs over the years and didn't squeal too much except when it really hurt. Second are the coteachers who used and/or contributed this material in working with me: Ken Boulding, who let me help out in his Senior Honors course at Michigan; Jim Greenwood, who took over for me at the IBM Systems Research Institute in New York; and Don Gause, who shares the teaching with me in the Human Sciences and Technology group at the State University of New York at Binghamton. Third are all those who taught me directly, but especially those to whom this volume is dedicated, Ken Boulding, Anatol Rapoport, and Ross Ashby. Fourth are all those whose material has been so liberally borrowed by me for the book, and may anyone who has not received proper credit please forgive me enough to let me know of my oversight. Fifth are the ever-so-many people who have contributed editorial work over the ever-so-many years this work has been in progress: especially Sheila Abend and Shanna and Mike McGoff. Finally, and in the place of greatest gratitude, are the two who read and ripped to shreds every word and diagram so as to convert a lumpy oatmeal pudding into what I hope is more like a wedding cake: Joan Kaufmann and Dani Weinberg.

Contents

1

The Problem

Today we preach that science is not science unless it is quantitative. We substitute correlation for causal studies, and physical equations for organic reasoning. Measurements and equations are supposed to sharpen thinking, but . . . they more often tend to make the thinking non-causal and fuzzy. They tend to become the object of scientific manipulation instead of auxiliary tests of crucial inferences.

Many—perhaps most—of the great issues of science are qualitative, not quantitative, even in physics and chemistry. Equations and measurements are useful when and only when they are related to proof; but proof or disproof comes first and is in fact strongest when it is absolutely convincing without any quantitative measurement.

Or to say it another way, you can catch phenomena in a logical box or in a mathematical box. The logical box is coarse but strong. The mathematical box is fine grained but flimsy. The mathematical box is a beautiful way of wrapping up a problem, but it will not hold the phenomena unless they have been caught in a logical box to begin with.

John R. Platt[1]

The Complexity of the World

It isn't what we don't know that gives us trouble, it's what we know that ain't so.

Will Rogers

The first step to knowledge is the confession of ignorance. We know far, far less about our world than most of us care to confess. Yet confess we must, for the evidences of our ignorance are beginning to mount, and their scale is too large to be ignored!

If it had been possible to photograph the earth from a satellite 150 or 200 years ago, one of the conspicuous features of the planet would have been a belt of green extending 10 degrees or more north and south of the Equator. This green zone was the wet evergreen tropical forest, more commonly known as the tropical rain forest. Two centuries ago it stretched almost unbroken over the lowlands of the humid Tropics of Central and South America, Africa, Southeast Asia and the islands of Indonesia.

  . . . the tropical rain forest is one of the most ancient ecosystems . . . it has existed continuously since the Cretaceous period, which ended more than 60 million years ago.

  Today, however, the rain forest, like most other natural ecosystems, is rapidly changing. . . . It is likely that by the end of this century very little will remain.[2]

This account may be taken as typical of hundreds filling our books, journals, and newspapers. Will the change be for good or evil? Of that, we can say nothing—that is precisely the problem. The problem is not change itself, for change is ubiquitous. Neither is the problem in the man-made origin of the change, for it is in the nature of man to change his environment. Man's reordering of the face of the globe will cease only when man himself ceases.

The ancient history of our planet is brimful of stories of those who have ceased to exist, and many of these stories carry the same plot: Those who live by the sword, die by the sword. The very source of success, when carried past a reasonable point, carries the poison of death. In man, success comes from the power that knowledge gives to alter the environment. The problem is to bring that power under control.

In ages past, the knowledge came very slowly, and one man in his life was not likely to see much change other than that wrought by nature. The controlled incorporation of arsenic into copper to make bronze took several thousand years to develop; the substitution of tin for the more dangerous arsenic took another thousand or two. In our modern age, laboratories turn out an alloy a day, or more, with properties made to order. The alloying of metals led to the rise and fall of civilizations, but the changes were too slow to be appreciated. A truer blade meant victory over the invaders, but changes were local and slow enough to be absorbed by a million tiny adjustments without destroying the species. With an alloy a day, we can no longer be sure.

Science and engineering have been the catalysts for the un-precedented speed and magnitude of change. The physicist shows us how to harness the power of the nucleus; the chemist shows us how to increase the quantity of our food; the geneticist shows us how to improve the quality of our children. But science and engineering have been unable to keep pace with the second-order effects produced by their first-order victories. The excess heat from the nuclear generator alters the spawning pattern of fish, and, before adjustments can be made, other species have produced irreversible changes in the ecology of the river and its borders. The pesticide eliminates one insect only to the advantage of others that may be worse, or the herbicide clears the rain forest for farming, but the resulting soil changes make the land less productive than it was before. And of what we are doing to our progeny, we still have only ghastly hints.

Some have said that the general systems movement was born out of the failures of science, but it would be more accurate to say that the general systems approach is needed because science has been such a success. Science and technology have colonized the planet, and nothing in our lives is untouched. In this changing, they have revealed a complexity with which they are not prepared to deal. The general systems movement has taken up the task of helping scientists to unravel complexity, technologists to master it, and others to learn to live with it.

In this book, we begin the task of introducing general systems thinking to those audiences. Because general systems is a child of science, we shall start by examining science from a general systems point of view. Thus prepared, we shall try to give an overview of what the general systems approach is, in relation to science. Then we begin the task in earnest by devoting ourselves to many questions of observation and experiment in a much wider context. And then, having laboriously purged our minds and hearts of "things we know that ain't so," we shall be ready to map out our future general systems tasks, tasks whose elaboration lies beyond the scope of this small book.

Mechanism and Mechanics

Physics does not endeavor to explain nature. In fact, the great success of physics is due to a restriction of its objectives: it endeavors to explain the regularities in the behavior of objects. This renunciation of the broader aim, and the specification of the domain for which an explanation can be sought, now appears to us an obvious necessity. In fact, the specification of the explainable may have been the greatest discovery of physics so far.

The regularities in the phenomena which physical science endeavors to uncover are called the laws of nature. The name is actually very appropriate. Just as legal laws regulate actions and behavior under certain conditions but do not try to regulate all actions and behavior, the laws of physics also determine the behavior of its objects of interest only under certain well-defined conditions but leave much freedom otherwise.[3]

Eugene P. Wigner

To understand the general systems view of science, we should examine physics—and particularly mechanics—because these sciences are often taken as the ideal of others. The beauty of the mechanical model of the world was well expressed by Karl Deutsch,[4] who said that mechanism

> . . . implied the notion of a whole which was completely equal to the sum of its parts; which could be run in reverse; and which would behave in exactly identical fashion no matter how often these parts were disassembled and put together again, and irrespective of the sequence in which the disassembling or reassembling would take place. It implied consequently that the parts were never significantly modified by each other, nor by their own past, and that each part once placed in its appropriate position with its appropriate momentum, would stay exactly there and continue to fulfill its completely and uniquely determined function.

The luster of this description is dulled a bit by the observation that mechanical systems ordinarily have but a handful of identifiable parts—most often 2, sometimes 10, or perhaps as many as 30 or 40 if they are highly constrained, as are the parts of a bridge. If there are too many parts, the physicist may write down equations relating the behaviors of the different parts, but he cannot solve the equations, even by approximate methods. True, high-speed computers have extended the range of approximate solutions to mechanical systems, but only by a relatively small amount.

If the formal methods of mechanics are so limited, why is mechanics considered a model for the sciences? We must—if we are to have the answer—consider not the formal methods, but the informal. Complex mechanical systems are *always* informally reduced to simpler ones. Only then can the formal methods begin to do their work.

Consider, for example, Newton's explanation of the motions of the bodies in the solar system. Rapoport,[5] in speaking about this problem, pointed out that

> Fortunately for the success of the mechanistic method, the solar system . . . constituted a special tractable case of several bodies in motion.

Although Rapoport's analysis is correct and to the point as far as it goes, it does not penetrate deeply enough into the heart of Newton's success. The solar system, in the first place, does not consist of "several bodies in motion." We now know that there are thousands upon thousands of celestial bodies in our solar system plus other matter not in "bodies." (See Figure 1.1.) Any analysis of planetary motions, however, begins by ignoring most of these bodies. They are said to be "too small" to have a significant effect on the calculations. (See Figure 1.2.) Although this seems a natural step—so natural that texts on mechanics do not ordinarily mention it—it happens to work only in very special circumstances. Any other circumstances are not considered proper systems for mechanistic thinking.

Figure 1.1.   There are thousands upon thousands of celestial bodies . . .

Consider, for instance, the pineal gland, a tiny piece of tissue in the brain. Can physiologists ignore this body in their attempts to understand the behavior of the human body? Perhaps they can—the question is quite alive—and perhaps they cannot. In any case, no physiologist would think of arguing that because the mass of the pineal gland is small with respect to the mass of the brain, it can be ignored on that account. The DNA in a living cell is a miniscule amount of the cell material, if measured according to mass; but understanding of cellular biology would be impossible without considering its role. The queen bee in a hive is only one of thousands of bees and makes up only a small fraction of the total mass of the hive, but no ethologist dare ignore her.

Mechanics, then, *is the study of those systems for which the approximations of mechanics work successfully.* It is strictly a matter of empirical evidence, not of theory, that the human body cannot be understood by considering only the gravitational attractions between its parts.

Figure 1.2. The analysis of planetary motions begins by ignoring most of these bodies . . .

The Square Law of Computation

Whereas in the past the only resource for dealing with biological systems was to try to minimize the interactions between the parts, thereby often losing the real focus of interest, today nothing but time and money prevent us from treating real biological systems in all their complexity and richness.[6]

W. Ross Ashby

What is the cost of computation, in time and in money? How important was ignoring small bodies (the asteroids, comets, satellites and other pieces of space flotsam) to the economical calculation of planetary orbits?

Consider first the equations needed to describe the most general system of only two objects. We must first describe how each object behaves by itself—the "isolated" behavior. We must also consider how the behavior of each body affects that of the other—the "interaction." Finally, we must consider how things will behave if neither of the bodies is present—the "field" equation. Altogether, the most general two-body system requires four equations: two "isolated" equations, one "interaction" equation, and one "field" equation.

As the number of bodies increases, there remains but a single "field" equation, and only one "isolated" equation per body. The number of "interaction" equations, however, grows magnificently, with the result that for n bodies we would need 2^n relationships! (See Appendix I, under "Scientific Notation," for an explanation of these exponential numbers.)

To be more concrete, for 10 bodies we would need $2^{10} = 1024$ equations and for 100,000 bodies, about $10^{30,000}$. By "ignoring small masses," then, the number of equations is reduced from perhaps $10^{30,000}$ to approximately 1000. At least it would now be possible to write down the equations, even if we still could not afford to solve them.

How much effort is involved in solving equations, and why are we so interested in the question? In Newton's day, the impact of mechanics on philosophical thought was pervasive. Many philosophers thought, with Laplace, that given precise observations on the position and velocity of every particle of matter, one could calculate the entire future of the universe. Although they realized that they would need a large computing machine, they lacked even the smallest computers. How could they possibly put a measure on the required computation?

Only in our lifetime have the dreams of the mechanists been realized, but with the realization came a revolution in philosophical thought. One aspect of this revolution was the more realistic concern for the question of computational cost, a question raised by the systems thinkers, but most notably and consistently by Ashby. This annoying question—*how much* "time and money"?—lies at the very foundation of the general systems movement.

We do not need exact measures. Instead, we merely want to estimate *how the amount of computation increases as the size of the problem increases.* Experience has shown that *unless some simplifications can be made,* the amount of computation involved increases at least as fast as the square of the number of equations. This we call the "Square Law of Computation." Thus, if we double the number of equations, we shall have to find a computer four times as powerful to solve them in the same amount of time. Naturally, the time often goes up faster than this—particularly if some technical difficulty arises, such as a decrease in the precision of results. For our present arguments, however, we may conservatively use the Square Law of Computation to estimate how much more computing is required for one general set of equations than for another.

In practice, then, there is an upper limit to the size of the system of equations that can be solved. Clearly, $10^{30,000}$ equations are far beyond that limit. And in Newton's day, without computers at all, the practical limit of computations was well below 1000 second-order differential equations, especially since Newton had just invented differential equations. Newton needed all the simplifying assumptions, explicit or implicit, he could get away with, just as physiologists and psychologists do today. We may note, in this regard, that old-time physicists now say that the "youngsters" no longer do "real physics." These young upstarts use the computer to solve large sets of equations, rather than applying physical "intuition" to reduce the equations so they can be solved with a pencil on the back of the proverbial envelope.

The Simplification of Science and the Science of Simplification

I do not know how it is with you, but for myself I generally give up at the outset. The simplest problems which come up from day to day seem to me quite unanswerable as soon as I try to get below the surface.

Justice Learned Hand

Thinking about the practical problem of computation, then, can give us a new point of view about what mechanics, or any science, is. Since practical computation demands that implicit assumptions be brought out into the open, it is no coincidence that computer programmers are attracted to an approach devoted to studying how people make assumptions. As an example of such computing experience, consider another assumption already made in our reduction of the solar system problem to 1000 equations.

We have assumed—as one always assumes in mechanics—that only certain *kinds* of interactions are important. In this case, the only important interaction was gravitational, which meant that each relationship gave only one equation. How do we know that only gravitational attraction is important in this system? How do we know that we can ignore magnetic effects, electrostatic forces, light pressure, force of personality, and so forth? One answer is that this problem would not be a problem in mechanics if other kinds of forces were important; but that answer is merely begging the question. *How do we know it is a problem in mechanics?*

As before, we know it is a problem in mechanics because when we try these approximations, they give us satisfactory answers, that is, answers that match observational data. If we had a problem for which they did not work, it would never make its way into the mechanics textbook. Our practical computing example of this quandary is the calculations that were made of the orbit of the Echo satellite, which was a large inflated Mylar sphere. The classical solution of the gravitational equations was not doing a satisfactory job of predicting Echo's orbit. After much perplexing labor, the programmers realized that because of its small density, Echo was much larger than any "normal" solar body of the same mass. Consequently, the pressure of the sun's light radiating upon its surface could not be implicitly ignored, as it is in all "ordinary" orbital calculations. No, mechanics does not tell us which systems are "mechanical."

And yet, even having reduced the number of equations to 1000—by applying deeply buried assumptions—we still may not be able to say we have solved this mechanical system. The equations may still prove intractable, even for a large computer. We need further simplifications. Newton supplied an important one in his Law of Universal Gravitation, which has been called "the greatest generalization achieved by the human mind."[7]

The law states that the force of attraction (*F*) between two (point) masses was given by the equation:

$$F = \frac{GMm}{r^2}$$

where *M* is the mass of the first, *m* is the mass of the second, *r* is the distance between them, and *G* is a universal constant. From the viewpoint of simplification, this equation says more implicitly than explicitly; for it states that *no other equation is needed*. It says, for instance, that the force of attraction between two bodies is in no way dependent on the presence of a third body, so that only pairs of bodies need be considered in turn, and then all of their effects may be added up (Figure 1.3).

A psychologist, for one, would be tickled pink if he could consider only summed pair interactions. This simplification would mean that, to understand the behavior of a family of three, he would study the behavior of the father and mother together, the father and son together, and the mother and son together. When all three interacted, their behavior could be predicted by summing their pairwise behaviors. Unfortunately, it is only in mechanics and a few other sciences that superposition of pairwise interactions can be successful.

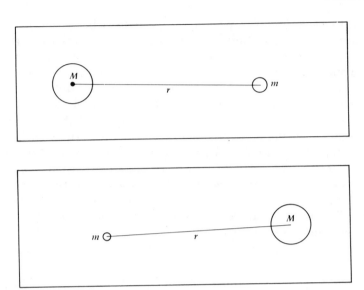

Figure 1.3. Only pairs need be considered in turn . . .

In the case of the solar system, pairwise superposition reduces 1000 equations to about 45, that being the number of ways 10 things can be taken in pairs. From a computational point of view, we have reduced the size of our task by the square of 1000/45 or about 100 times, at least. We might be willing to stop at this point, although Newton, perhaps because he did not have the computers we have, went still further.

As it happens, the solar system has one body (the sun) whose mass is much larger than any of the other masses, larger, in fact, than the mass of all of the other bodies together. Because of this dominant mass, the pair equations not involving the sun's mass yield forces small enough to be ignored, at least considering the accuracy of the data Newton was trying to explain. (Discrepancies in this assumption led to the discovery of at least one planet that Newton did not know.) This simplification, which is made possible by the solar system, rather than by mechanics, reduces the number of equations to about 10, instead of 45, giving an estimated $20\times$ reduction in computation.

But Newton went even further than this, for he observed that the dominant mass of the sun enabled him to consider each planet together with the sun as a separate system from each of the others. Such a separation of a system into noninteracting subsystems is an extremely important technique known to all developed sciences—and to systems theorists as well. To understand the power of such a separation, we need only recall the Square Law of Computation. If solving a system of n equations takes n^2 units of computation, n separate single equations taken one at a time will take only n of the same units (Figure 1.4).

At this point, Newton stopped simplifying and solved the equations analytically. He had actually made numerous other simplifications, such as his consideration of each of the solar bodies as point masses. In each of these cases, he and his contemporaries were generally more aware of—and more concerned about—the simplifying assumptions than are many present-day physics professors who lecture about

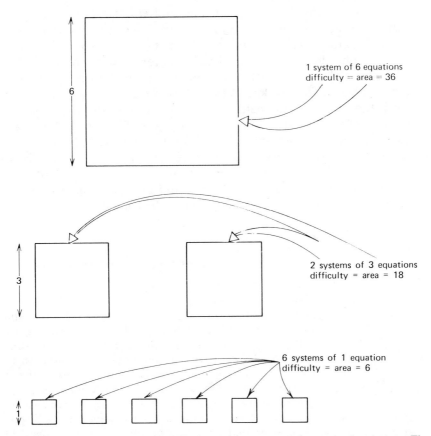

1 system of 6 equations
difficulty = area = 36

2 systems of 3 equations
difficulty = area = 18

6 systems of 1 equation
difficulty = area = 6

Figure 1.4. The power of separation. Each square represents a set of equations. The side of the square represents the number of equations n. The *area* of the square then represents the difficulty of computation n^2. By dividing the 6 equations into 2 sets of 3, we reduce the area of the squares from 36 to 18. By dividing the 6 equations into 6 sets of 1, we reduce the area of the squares from 36 to 6.

Newton's calculations. Students, consequently, find it hard to understand why Newton's calculation of planetary orbits is ranked as one of the highest achievements of the human mind.

But the general systems thinker understands. He understands because it is his chosen task to understand the simplifying assumptions of a science—in Wigner's words, those "objects of interest" and "well-defined conditions" *that delimit its domain of application and magnify its power of prediction.* He wants to go right to the beginning of the process by which a scientist forms his models of the world, and to follow that process just as far as it will help him in suggesting useful models for other sciences.

Why is the general systems thinker interested in the simplifications of science—in the science of simplifications? For exactly the same reason as Newton was. The systems theorist knows that the Square Law of Computation puts a limit on the power of any computing device. Moreover, he believes that the human brain is in some sense a computing device. Thus he knows that, if we are to survive in this complex world, we need all the help we can get. Newton was a genius, but not because of the superior computational power of his brain. Newton's genius was, on the contrary, his ability to simplify, idealize, and streamline the world so that it became, in some measure, tractable to the brains of perfectly ordinary men. By studying the methods of simplification that have succeeded and failed in the past, we hope to make the progress of human knowledge a little less dependent on genius.

Statistical Mechanics and the Law of Large Numbers

In the space of one hundred and seventy-six years the Lower Mississippi has shortened itself two hundred and forty-two miles. That is an average of a trifle over one mile and a third per year. Therefore, any calm person, who is not blind or idiotic, can see that in the Old Silurian Period, just a million years ago next November, the Lower Mississippi River was upward of one million three hundred thousand miles long, and stuck out over the Gulf of Mexico like a fishing rod. And by the same token any person can see that seven hundred and forty-two years from now the Lower Mississippi will be only a mile and three-quarters long, and Cairo and New Orleans will have joined their streets together, and be plodding along under a single mayor and mutual board of aldermen. There is something fascinating about science. One gets such wholesome returns of conjecture out of such a trifling investment of fact.

Mark Twain, *Life on the Mississippi*

Newton's achievement was in describing the behavior of a system of perhaps 10^5 objects, of which he found 10 of interest. By the nineteenth century, however, physicists wanted to tackle other systems, simple little systems such as the molecules in a bottle of air.

The molecules in a bottle of air differ from the solar system in several ways. First of all, there are not 10^5 of them, but 10^{23}. Second, the nineteenth-century physicists were not interested in just 10 of the molecules, but in all of them. Third, had they been interested in only 10, they would have had to study all 10^{23}, since the molecules were pretty much identical in mass and were, furthermore, in close interaction (Figure 1.5).

These nineteenth-century physicists already knew from Newton that they only had to consider pair relations, but this merely reduced the number of equations from about $2^{10^{23}}$ to 10^{46}. Although this is undoubtedly a substantial reduction, the prospects of further reduction of that 10^{46} looked rather grim. After a few fruitless tries at the job, these

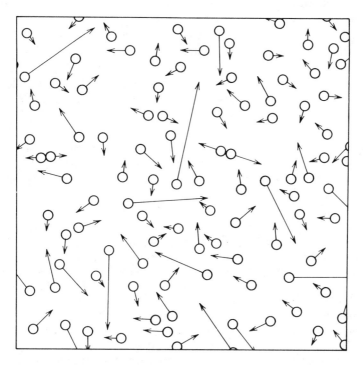

Figure 1.5. There were 10^{23} molecules, identical in mass, and in close interaction.

physicists must have felt much like the fox in Aesop's fable who just could not quite reach the grapes. We know they must have felt that way because they solved their problem the same way the fox did: They decided that they did not really want to know about the individual molecules anyway.

Actually, of course, the matter was not entirely one of sour molecules. We might more realistically describe the position of these physicists (such as Gibbs, Boltzmann, and Maxwell) by saying that they were *lucky* not to be interested in things for which they could not solve their equations. They had inherited a set of observed laws (such as Boyle's Law) about the behavior of certain measurable properties of gases (such as pressure, temperature, and volume). They believed that gases were made of molecules, but they had to bridge the gap between that belief and the observed properties of gases. They bridged that gap by postulating that the interesting measurements were a few *average* properties of the molecules, rather than the exact properties of any one molecule (Figure 1.6).

Since the number of different average properties was small, this simplification brought down the amount of computation in one fell swoop. Furthermore, the precision of prediction that was obtained for the averages was excellent, because the number of molecules was very, very large, and therefore the so-called "Law of Large Numbers" could be invoked. What this law says, in essence, is that the larger the population, the more likely we are to *observe* values that are close to the *predicted* average values.

More precise statements of the Law of Large Numbers enable us to say just how close we may expect observed and predicted values to be, depending on the size of the population. The most useful rule of thumb (or general systems law) in this context is Schrödinger's "Square Root of N Law":

If I tell you that a certain gas under certain conditions of pressure and temperature has a certain density, and if I expressed this by saying that within a certain volume (of a size relevant for some experiment) there are under these conditions just n molecules of the gas, then you might be sure that if you could test my statement in a particular moment of time, you would find it inaccurate, the departure being of the order of $\sqrt{n}$. Hence if the number $n = 100$, you would find a departure of about 10, thus relative error = 10%. But if $n =$ 1 million, you would be likely to find a departure of about 1000, thus relative error = 1/10%. Now, roughly speaking, this statistical law is quite general. The laws of physics and physical chemistry are inaccurate within a probable relative error of the order of $1/\sqrt{n}$, where n is the number of molecules that co-operate to bring about that law—to produce its validity within such regions of space or time (or both) that matter, for some considerations or for some particular experiment.

You see from this again that an organism must have a comparatively gross structure in order to enjoy the benefit of fairly accurate laws, both for its internal life and for its interplay with the external world. For otherwise the number of co-operating particles would be too small, the "law" too inaccurate. The particularly exigent demand is the square root. For though a million is a reasonably large number, an accuracy of just 1 in 1000 is not overwhelmingly good, if a thing claims the dignity of being a "Law of Nature."[8]

(See Figure 1.7.) In this vivid passage, Schrödinger not only explains why the laws of physics and physical chemistry work so well; but he

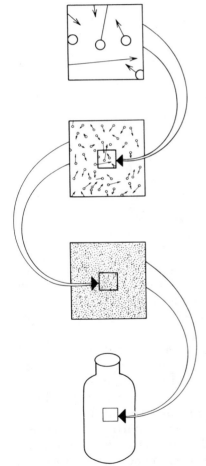

Figure 1.6. The interesting measurements were only a few *average* properties.

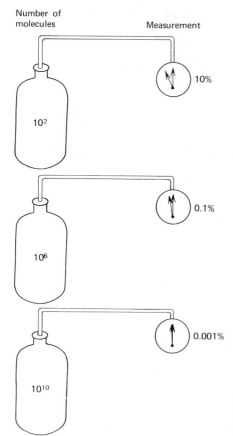

Figure 1.7. The departure is of the order of the square root of the number of molecules.

goes on to explain a design principle that organisms should follow if they too are to "enjoy the benefit of fairly accurate laws." For now, however, we are only interested in the usefulness and limitations of the statistical approach to problems in other fields of science and technology.

What is the scope of the statistical approach? How does it relate to the scope of the purely mechanical? One suggestive phrase is that statistical mechanics deals with "unorganized complexity"—that is, systems that are complex, but yet *sufficiently random* in their behavior so that they are *sufficiently regular* to be studied statistically.

The concept of "randomness" is most important for systems thinking, though randomness often leads to properties quite contrary to our intuition. We do not have such a problem in understanding the success of mechanics, for although "simplicity" will prove to be as slippery a concept as "randomness," to a first approximation we were able to use the number of objects as a measure of complexity—the complement of simplicity.

Intuitively, *randomness is the property that makes statistical calculations come out right.* Although this definition is patently circular, it does help us to understand the scope of statistical methods. Consider a typical statistical problem. There is a flu epidemic and we want to know how it will spread through the population so that we may plan for the distribution of vaccine. If every person is just as likely to get the disease as any other, we can calculate the expected number of cases and the effect of vaccination strategies with great precision. If, on the other hand, there is some sort of nonrandomness in the population, our simple calculations will begin to deviate from the experienced epidemic.

What could be a source of such nonrandomness? To take one case, people are not randomly distributed around the countryside, so that

= Analytical treatment

= Statistical treatment

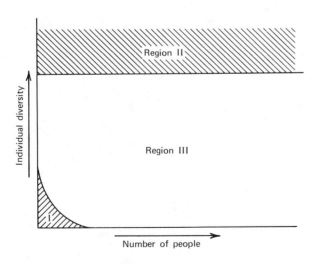

Figure 1.8. Types of populations in the prediction of epidemics.

the chances of exposure are not the same for everyone. If it were a simple (small) population, we could calculate the exact exposures for every member, but we use a statistical approach just because the population is not small. In a small population, the very knowledge of the precise nature of the interactions would be what we needed in order to calculate the pattern of infection. On the other hand, in the large population, we have already given up hope of calculating the *exact* pattern and want to calculate *averages,* which are deranged by underlying structure. Thus the very type of structure that helped us in one approach hinders us in the other.

It may assist in understanding the situation to conceptualize it as shown in Figure 1.8. There we consider the size of each possible population as one axis of a plot and the "diversity" as the other. In the lower left corner (Region I) are small populations with a great deal of structure, which, as the chart shows, could be treated analytically. At the top, above the straight line in Region II, we have sufficient

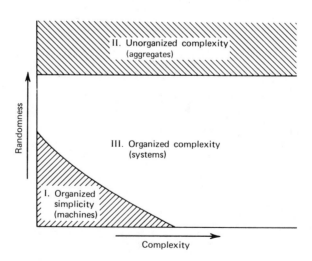

Figure 1.9. Types of systems with respect to methods of thinking.

diversity or randomness to achieve some desired precision of prediction. In Region III, in between, lie all the populations that are too diverse for analysis and too structured (perhaps because they are too small) for statistics.

Passing from this specific example to a more general case, we may obtain the chart shown in Figure 1.9, which is a relabeled version of Figure 1.8. "Number of people" has been generalized to "complexity", and "individual diversity" has been generalized to "randomness." In keeping with the very rough nature of the arguments, we do not put any numbers on the chart, but only concern ourselves with its general characteristics.

Region I is the region that might be called "organized simplicity"—the region of *machines,* or *mechanisms.* Region II is the region of "unorganized complexity"—the region of *populations,* or *aggregates,* as we shall call them. Region III, the yawning gap in the middle, is the region of "organized complexity"—the region too complex for analysis and too organized for statistics. This is the region of *systems.*

The Law of Medium Numbers

The mechanistic world view, taking the play of physical particles as ultimate reality, found its expression in a civilization which glorifies physical technology that has led eventually to the catastrophes of our time. Possibly the model of the world as a great organization can help to reinforce the sense of reverence for the living which we have almost lost in the last sanguinary decades of human history.[9]

Ludwig von Bertalanffy

Although technology often leads science in discovery, the philosophy of technology is usually drawn from the scientific philosophy of its time. In our time, the technology of machines has drawn its inspiration from mechanics, dealing with complexity by reducing the number of relevant parts. The technology of government, on the other hand, has drawn upon statistical mechanics, creating simplicity by dealing only with people in the structureless mass, as interchangeable units, and taking averages. As von Bertalanffy suggests, these philosophies may result from the lack of any scientific means of dealing with systems between these extremes—systems in the vast no-man's land of *medium numbers.*

For systems between the small and large number extremes, there is an essential failure of the two classical methods. On the one hand, the

Square Law of Computation says that we cannot solve medium number systems by analysis, while on the other hand, the Square Root of N Law warns us not to expect too much from averages. By combining these two laws, then, we get a third—the *Law of Medium Numbers*:

For medium number systems, we can expect that large fluctuations, irregularities, and discrepancy with any theory will occur more or less regularly.

The importance of the Law of Medium Numbers lies not in its power of prediction, but in the scope of its application. Although good mechanical and statistical systems are actually quite rare, we are literally surrounded by medium number systems. Computers have medium numbers of components, cells have medium numbers of enzymes, organizations have medium numbers of members, people have medium numbers of vocabulary words, and forests have medium numbers of trees, or flowers, or birds.

As with most general systems laws, we find a form of the Law of Medium Numbers in folklore. Translated into our daily experience—combining our familiarity with such systems and our ineptitude in their face—the Law of Medium Numbers becomes *Murphy's Law*:

Anything that can happen, will happen.

Science, like any of us, is unable to cope with medium number systems, though its success with systems of its own choosing has misled many scientists and politicians into thinking of science as a way of dealing effectively with *all* systems. Science, as science, is no more to blame for the consequences than a band saw is responsible for the consequences of its being used to trim fingernails. If fingernails need cutting, and the band saw is the only available cutting tool, then the results are more or less predictable. A band saw is a most useful tool, but not for certain jobs.

Science, too, is a most useful tool—probably the most useful that man has ever discovered. Not even the most ardent naturist would actually be willing to live without tasting any of the fruits of science. But the fruits of science are simple fruits, or more precisely, fruits of simplification. Social scientists, for example, study us as great masses of humanity in order to plan for our overall needs, and engineers satisfy those needs by putting together small numbers of parts into large machines whose essential principle is to keep the parts from strong interactions.

Many of the ills of society come from too good an application of these simple fruits: a richness of means applied to impoverished ends. Yet many more—and perhaps the impoverished ends themselves—come from a vastly inadequate technology attempting what it can never do. We must begin to understand the limitations of brash technology, for its principal method is to squelch medium number systems.

Consider, for example, the methods of simplification applied to a large electronic device, such as a computer. The individual transistors depend on a single physical law and are manufactured with the utmost purity so that the law will hold within them. Such devices, though there may be 100,000 of them in one unit, rarely cause trouble. On the other hand, troubles frequently arise in the joints by which the transistors are connected to each other and to the rest of the system. Why? Because the purity of the transistor is achieved by pushing out to the joints all the dirty problems of physical strength, exposure to air, bonding of unlike materials, and so forth.

This separation of function has been carried to great extremes in our technology. Whenever a designer can transcend it, he can create a whole new technology. From time to time it is recognized that some device is not merely a collection of components, but a collection together with the relationships (joints) between them. Then, a new level of device—the "integrated circuit," for example—is created in which the previous components lose their separate identities as the joints disappear. This new device, in turn, becomes a "component" in the new way of thinking, and the connections to it again become the weakest part of the system.

Separation of function is not to be despised, but neither should it be exalted. Separation is not an unbreakable law, but a convenience for overcoming inadequate human abilities, whether in science or engineering. As one of the spiritual fathers of the general systems movement said:

As we analyze a thing into its parts or into its properties, we tend to magnify these, to exaggerate their apparent independence, and to hide from ourselves (at least for a time) the essential integrity and individuality of the composite whole. We divided the body into its organs, the skeleton into its bones, as in very much the same fashion we make a subjective analysis of the mind, according to the teaching of psychology, into component factors: but we know very well that judgement and knowledge, courage or gentleness, love or fear, have no separate existence, but are somehow mere manifestations, or imaginary coefficients, of a most complex integral.[10]

The world is one whole. The fragmentation of knowledge about the world is exactly analogous to the separation of a device into its components, the body into its organs, or the surface of the earth into political units: It is useful in some situations but is usually carried to extremes. Eventually, we get revolutionary movements for new syntheses of knowledge, creating whole new fields, such as electro-magnetism, physical chemistry, social psychology, or maybe even psychobotany, or movements for a new political synthesis, creating new forms of economy, culture, and society.

The biological and social sciences are not as "successful" as the physical sciences. They are not so free to chop up the world as it is given to them, for the piece they have taken for their own is essentially indivisible. Anatomists have had some success, but we are not so interested in how a man operates when he is disassembled. Social scientists have had even less success, because their main interest—"hu-manness"—is a medium number property that disappears when the system is taken apart or abstracted and averaged. When behavioral scientists try to understand the "individual" by averaging, the properties of the individual are smoothed out and lost. When they try to isolate the individual for study, they disconnect their subject from other men and other parts of the world so that he becomes merely a laboratory artifact—and something less than human.

For most of his short history, man's physical environment was only indirectly and partially under his control. Very recently, man invented science to increase that control, and he has been so fascinated by the quick and easy success that he has not paid much attention to conse-quences outside his analyses and averages. As a result, we come to ex-pect that the future will see even greater mastery of the environment—and of man himself.

But all too frequently, that mastery seems to be accompanied by creeping slavery. Perhaps we are beginning to feel the results of treating a system as if it were a collection of parts, and an individual as if he were only contributing to an average. Perhaps, too, we are reaching the useful limits of a science and technology whose philo-sophical underpinnings are techniques restricted to systems of small and large numbers.

The general systems movement itself, of course, is subject to the same abuses if it carries its principles beyond their useful limits. General systems thinking is not going to yield the kinds of control over medium number systems that we might imagine we would like to have, and its major contribution is most likely to be in limiting the excesses of other approaches to complexity. Still, if we want to reverse the trend of "the last sanguinary decades of human history," we may have to turn more and more to some sort of synthesis. We already know how to transform prairies into dustbowls, lakes into cesspools, and cities into mausoleums. Can we turn around before it is too late?

QUESTIONS FOR FURTHER RESEARCH

1. *Economics*

Vilfredo Pareto, in his famous *Manuel d'Economie Politique,* mentioned that this general equilibrium theory, when applied to a system of 100 persons and 700 commodities, would require not less than 70,699 equations to be solved. Where does this figure come from, and how is it dependent on the number of persons and commodities? What does it mean for Pareto's theory? What might be done to save the theory from this vast number of equations.?

2. *Social Psychology and Sociology*

A frequently used method in studying group structures is the so-called sociometric method, perhaps modeled after the "econometric" method in economics. The method, originated by J. L. Moreno in his book, *Who Shall Survive?* (1934), has been elaborated in a number of directions by later workers. Essentially, the method involves determining the strength or quality of interactions between all pairs of persons involved in a group, possibly along several dimensions, such as like–dislike, interact–avoid, important–indifferent. What might be the limit on the size of the group that could be studied effectively with such methods? Could this limit be the dividing line between social psychology and sociology? Under what special circumstances could larger groups be studied with such techniques?

3. *Mechanics*

For those who doubt the degree with which the success of physics depends upon the reduction of complex systems to simple ones, we need only reflect upon the three-body problem. As soon as a third body is added to the completely solved pair of bodies, the solution, in general, becomes impossible. Whereas the two-body problem can be solved by a high-school student, the intractability of the three-body problem can be gauged by considering that in July 1969, an international gathering of physicists met in Birmingham, England, to consider the "Three-Body Problem in Nuclear and Particle Physics." The proceedings of this conference, while dealing only with special cases of the three-body problem, contained over 70 papers on the subject—and, of course, the problem is still unsolved. Those interested in the success of physics at dealing with complex systems should prepare a report summarizing this conference.

> *Reference:* J. S. C. McKee and P. M. Rolph, Eds., *Three Body Problems in Nulcear and Particle Physics,* Proceedings of an International Conference, Birmingham, England, July 1969. New York: Elsevier, 1970.

4. *Archaeology*

The complexity of seemingly simple objects is nowhere better illustrated than in archaeology, where from a single piece of stone that most of us would dismiss as uninteresting, whole patterns of a vanished society can be deduced. The collection:

> Martin Levey, Eds., *Archaeological Chemistry: A Symposium.* Philadelphia: University of Pennsylvania Press, 1967.

assembles 15 different reports on the extraction of information from small bits of matter as done by archaeologists. How does the work of these archaeologists compare with the work of the theoretical physicist? What simplifications do they share, and what simplifications dare they not share?

5. *Thermodynamics (or "Thermostatics")*

Of the three commonly observed states of matter, gases were the first to be reasonably well understood by physicists, starting, perhaps, with Boyle's Law. More recently, crystalline solids have become quite tractable for the physicist, but liquids remain the least known of the states. Discuss this historical sequence in terms of the Law of Medium Numbers.

6. *Operations Research*

The word "computation" in the Square Law of Computation does not necessarily refer to "solving equations" in only the ordinary senses of the term. A computer *simulation* is a method of computation in which the "equations" do not necessarily appear explicitly. Imagine that we were simulating a production line, perhaps the assembly line for automobiles or a petroleum distillery. What are the changes in the simulation that will tend to increase the computation by the Square Law? What elements might be found in the process that will permit more detailed modeling than the Square Law might suggest?

> *Reference:* Thomas H. Naylor et al., *Computer Simulation Techniques.* New York: Wiley, 1966.

7. *"Science" as Science*

The misnaming of fields of study is so common as to lead to what might be general systems laws. For example, Frank Harary once suggested the law that any field that had the word "science" in its name was guaranteed thereby not to be a science. He would cite as examples Military Science, Library Science, Political Science, Homemaking Science, Social Science, and Computer Science. Discuss the generality of this law, and possible reasons for its predictive power.

8. *Poetry*

Tagore said, "By plucking her petals you do not gather the beauty of the flower." Many poets are similarly renowned for their celebration of wholeness and complexity. Choose a particular poet and several representative works to discuss in the light of the Law of Medium Numbers.

9. *Neuroendocrinology*

Not many years ago, the pineal gland (then called pineal *body*) was thought by some anatomists to have no function, perhaps because of its small size. Today, the situation is reversed, with investigators attributing to this tiny piece of tissue actions on the midbrain, hypothalamus, and pituitary; participation in the syntheses of various enzymes and other vital substances; and modification of brain activity and behavior. Discuss the history of investigation of this organ in the light of scientific simplification.

> *Reference:* G. E. W. Wolstenholme and Julie Knight, Eds., *The Pineal Gland.* Baltimore: Williams and Wilkins, 1971.

10. *Utopian Thought*

The ingestion of current scientific philosophy into popular thought is nowhere better illustrated than in Utopian writings. The French philosopher Saint-Simon lived at the beginning of the nineteenth century and was the inspiration of many Utopians of that time. He worked before the rise of statistical mechanics, and completely under the spell of Newtonian mechanics—so much so that he had a vision that Newton, not the Pope, had been elected by God to transmit His divine plan to humanity. Saint-Simon was especially interested in a "law of Universal Gravitation" for social bodies, evidently to make the social system as harmonious as the solar system.

It is a fascinating exercise, full of unexpected side branches into history, to trace the evolution of Utopian thought as influenced by the dominant scientific philosophy of its time.

> *Reference:* Edmund Wilson, *To the Finland Station; A Study in the Writing and Acting of History.* New York: Harcourt Brace, 1940.

READINGS

RECOMMENDED

1. Ludwig von Bertalanffy, "The History and Status of General Systems Theory." In *Trends in General Systems* Theory. George J. Klir, Ed. New York: Wiley, 1972.
2. Karl Deutsch, "Mechanism, Organism, and Society." *Philosphy of Science,* **18,** 230 (1951).

SUGGESTED

1. Erwin Schrödinger, *What is Life?* Cambridge: Cambridge University Press, 1945.
2. Kenneth Boulding, *The Image.* Ann Arbor: University of Michigan Press, 1956.

2

The Approach*

Then what is the answer?—Not to be deluded by dreams.
To know that great civilizations have broken down into violence,
and their tyrants come, many times before.
When open violence appears, to avoid it with honor or choose
the least ugly faction; these evils are essential.
To keep one's own integrity, be merciful and uncorrupted and
not wish for evil; and not be duped
By dreams of universal justice or happiness. These dreams will
not be fulfilled.
To know this, and know that however ugly the parts appear
the whole remains beautiful. A severed hand
Is an ugly thing, and man disseevered from the earth and stars
and his history . . . for contemplation or in fact . . .
Often appears atrociously ugly. Integrity is wholeness,
the greatest beauty is
Organic wholeness, the wholeness of life and things, the divine
beauty of the universe. Love that, not man
Apart from that, or else you will share man's pitiful confusions,
or drown in despair when his days darken.[1]

Robinson Jeffers, "The Answer"

Organism, Analogy, and Vitalism

I replied to someone who said I didn't see women as I represent them: "If I met such women in life, I should run away in horror." First of all, I do not create a woman, I make a picture.[2]

Henri Matisse

* The main tone of this chapter was set in the early 1960s, when I was spending a lot of time with Kenneth Boulding. Ten years later, upon rereading his article,[10] I realize that I cannot separate my views from his. Since he is much smarter than I, one can safely assume that whatever is coincident between this chapter and that article is his. Since he is a far better writer than I, the reader is strongly advised to read the article—even if he skips this chapter.

The general systems movement attempts to aid thinking about medium number systems by finding general laws. Although these laws are stated informally to aid recall and initial understanding, an essential part of the general systems approach is the insistence that they be supportable, if necessary, by rigorous operations on rigorously defined models. This insistence is to some extent a reaction to the bad reputation of previous approaches to medium number systems, most of which can be classed as *organismic*. Faced with systems of organized complexity, some thinkers turned to living systems as models. Knowledge about living systems was applied by *analogy* to other systems, in order to gain some point of leverage on the complexity.

Hobbes, for instance, viewed the state, the "body politic" as a literal body of a giant person, with the various organs representing the government agencies. Lamarck endowed plants and animals with a kind of "intelligence" by which they directed their evolution (Figure 2.1). Such analogies suffered first of all from lack of real knowledge of the analog. Hobbes' knowledge of physiology was inaccurate and incomplete, so how could he hope to draw useful conclusions from likening the state to a person? Lamarck certainly did not understand "intelligence" any better than we do today, so what was gained by modeling evolution on it? Today, in fact, most of the modeling actually goes the other way.[3]

Actually, sometimes we can gain from modeling on a poorly known system. A fresh point of view can be helpful if *something* is known about the analog. At the very least, an analog jiggles the mind—and heaven knows our minds need a little jiggling. In point of fact, however, the organismic analogies were not all that careful, and the approach suffered as much from poorly drawn analogies as from poorly understood analogs. By requiring the possibility of making the models rigorous, systems thinkers hope to avoid these pitfalls.

Without this possibility of rigor, it is too tempting to ignore discordant parts of the model—something for which the organismists were severely chastised. Again, however, this might not be too serious a practice as long as we do not take our models too seriously. We do not create the world, we make a model. That is what organismists do; that is what painters do; and that, too, is what scientists do, despite any protestations to the contrary.

Every model is ultimately the expression of *one thing we think we hope to understand in terms of another that we think we do understand.* The chain of reasoning may be a hundred logical steps or a single analogical leap, but always ends in terms of some primitives that we agree among ourselves not to question further. For a science to have explana-

Figure 2.1. Lamarck endowed plants with a kind of "intelligence" by which they directed their evolution.

tory "power," this set of primitives must be neither too large nor too small. For instance, animistic religions explain the behavior of every object by referring to its unique spirit. If a tree falls, that is the spirit of the tree. If a rock fails to move, that is the spirit of the rock. In Western religions, such explanations are not satisfying.

Some of us are more satisfied with reducing everything to a single primitive. If a tree falls, God willed it to fall. If a rock fails to move,

God willed it not to move. But if something explains everything, it explains nothing. That, at least, is the *scientific* view, which is why organismic theories got into trouble with scientists.

The mechanists claimed that every phenomenon could be reduced to the primitives of physics, or physics and chemistry. They did not actually demonstrate this for "everything," but only claimed it. Some organismists countered by saying that not everything could be reduced to those primitives—the process for living systems had to stop sometime earlier, with something called the "life force," or "vital essence." While "vital essence" is no more intrinsically mysterious than, say, "mass," *everything* the organismists did not understand was reduced to vital essence, which meant that vital essence really explained nothing, because, like God, it explained everything.

Whatever else such an explanatory primitive does, it *discourages scientific investigation.* Science is the study of those things that can be reduced to the study of other things. Science, in other words, is *essentially* reductionist, and the reductionists were right in saying that the vitalists were unscientific. When reductionists began to back their claims by reducing some of the "organismic" phenomena to physical or chemical primitives, the backers of "organism" went into hasty retreat.

It should be noted, however, that the reductionists have not yet succeeded in reducing all phenomena to physical and chemical primitives. Whether they can or not is a neat philosophical question, not a scientific one. The fact remains that there are lots and lots of medium number systems—not all of them "living systems" by any means—that have not been "explained" in terms of physical and chemical primitives. Among those systems are quite a few we cannot simply ignore while waiting for the reductionists to reduce.

Reductionism, in the end, is an article of faith, one that drives scientists to carry out certain investigations in the faith that they will thereby better "understand." No physicist, however, actually lives his life by his reductionist creed. He does not decide what he will eat for supper based on the length–mass–time–charge– · · · reduction of the items on the menu. He employs other primitives, such as the smells–looks–costs– · · · system of units. Neither does he, when at work in the laboratory, rely on his most primitive of primitives. He uses a particular support because it is "strong," or a particular meter because it is "stable." In doing so, he is making use of animist analogs, but he may still be doing excellent physics.

What we are saying is that the baby of organismic thinking should not be thrown out with the bath of vital essence. Vitalism is not a prescription for thinking, but the very opposite—a declaration that certain things are not to be thought about further. Organismic thought, on the other hand, is simply the reliance on analogy, something that every physicist has done from Newton forward and back. Every significant thinker in science has drawn upon useful analogies for simplifying certain stages of thought. What is important is not to *stop* with rough analogy when the occasion demands that we go on, but to render the analogy into a precise, explicit, and predictive model.

In our time, biological systems are much better understood than they were a century ago; therefore organismic analogies may now prove more fruitful. The general systems approach, however, need not limit itself to organismic analogies. To the extent that we can reduce the models of a science to explicit form, we can make models in any other field by analogy—but an analogy with known mathematical characteristics. Therefore, we want to understand and communicate the ways in which thinkers in all fields use analogies and, when necessary, convert them into models.

The Scientist and His Categories

Man is by nature metaphysical and proud. He has gone so far as to think that the idealistic creations of his mind, which correspond to his feelings, also represent reality.[4]

Claude Bernard

If we are to try and discover what the thinking of different disciplines has in common, we shall have to raise many epistemological questions—"How do we know what we know?" We shall not, however, take a philosophical approach to this subject, but a practical one. That is, we shall not ask "How do we know that what we know is true?" but instead, "How do we come to hold the ideas that we hold as knowledge" We are interested, in other words, in *how thinking is done,* not in proving that thinking is correct or incorrect.

Much thinking is done in completely personal, idiosyncratic terms, so much so that how it is done is incommunicable. There exist, however, many overt categories of thought, and many others that can be brought to the surface by a modest effort at introspection. Since our interest, fortunately, lies closer to "the conceptual schemes of science" than to "the delusions of psychotics," we may rely on a certain measure of public behavior to study.

Among the scientists, the anthropologists come closest to doing our kind of work when they study the conceptual schemes of naturally evolving social groups. Conceptual schemes are also found, however, in any subculture that develops when people work together. By possessing a common set of standard categories of thought—usually symbolized by special words or phrases—groups can simplify the process of *internal* communication. Paradoxically, the more effective these categories are for internal communication, the more difficult they make communication with outsiders.

The anthropologist is faced with this problem in an obvious way when she endeavors to become a "participant–observer" in some culture.* To become a participant–observer, one must first become a participant, which involves at a minimum learning the native language—and actually involves far, far more learning of nonlinguistic patterns.[5] In the same way, becoming part of a working subculture means learning to use its forms of thought and communication.

In modern industrial society, most of us operate within the realms of a variety of groups and so have learned not only subcultural patterns, but also how to switch effortlessly from one pattern to another. The physicist generally experiences no difficulty in switching from the language of celestial mechanics to the language of auto mechanics. Only when he *does* experience difficulty is he likely to remark that category schemes seem to hinder communication. Moreover, when he does remark, he will ordinarily identify the "foreign" language of the auto mechanic as being the source of the difficulty. The anthropologists call this bias "ethnocentrism."

One manifestation of ethnocentrism is the belief that one's own culture is "superior" to those that one does not understand, or, rather, one whose natives "don't understand us, even though we are speaking perfectly clear English." If we but carry the "white man's burden" to the natives, surely they will make us their leader, or their god. After all, "in the kingdom of the blind, the one-eyed man is king."

Well, it never turns out quite that way, as H. G. Wells knew when he wrote his story "The Country of the Blind." A one-eyed man who happens upon a blind society does not become its king, for he cannot even function and is thought to be insane or sick. Because of the importance of category systems in a social group, it is not the outsider with a "better" system that becomes king, but the insider who most thoroughly masters the internal system. Should one of these "leaders" be removed to another group, his "native talent" evaporates, and probably becomes a severe handicap.

* If the "she" in this sentence gave you a little surprise, one of your category systems is showing.

On the other hand, there are people who experience difficulty fitting into their own native group, yet who always seem to be moderately successful at getting by anywhere else. The anthropologist, for one, tends to be this way. Although a professional at fitting into all sorts of exotic cultures, when he comes home he never becomes smoothly integrated: a critic at home and a conformist elsewhere.

The disciplines within science also form social groups, and thus have category schemes to facilitate internal communication. Thomas Kuhn, in *The Structure of Scientific Revolutions,*[6] has begun the study of the ways in which new "paradigms" are created and old ones destroyed; how paradigms are transmitted from one generation to the next; and how paradigms both help and hinder the progress of science. In particular, he distinguished between "normal science"—working *within* the current paradigms, and "scientific revolutions"—in which the paradigms themselves come under assault.

If our observations about category schemes generalize to sciences, then "leading scientists" should be the least likely people to lead scientific revolutions. Kuhn concurs in this conclusion, as did Max Planck in his *Scientific Autobiography*[7]:

A new scientific truth does not triumph by convincing its opponents and making them see the light, but rather because its opponents eventually die, and a new generation grows up that is familiar with it.

The average scientist is good for at most one revolution. Even if he has the power to make one change in his category system and carry others along, success will make him a recognized leader, with little to gain from another revolution.

Paradoxically, some scientists lead revolutions in several *different* disciplines—though not through a change in personal category systems. On the contrary, they carry a paradigm intact from one discipline to the other. While the colonialist may not master the natives' category system, he may become their king by introducing an entirely new element. One need not learn the language if one has the only rifle in town, lots of bullets, and the will to do a little shooting.

When England built the Empire, the young men who carried the white man's burden were seldom successes at home. They may have had talent, but the rather rigid social system had no room for them. In the same way, paradigms are often carried across disciplines by people who are having trouble rising in their own disciplinary hierarchy, whether for lack of talent or lack of room. In general, however, these "interdisciplinarians" are not what we would call "generalists." Like the mole, they know one thing exceedingly well, and apply it again and again to whatever discipline will have them.

The generalist, on the other hand, is like the fox, who knows many things. Just as anthropologists learn to live in many cultures, without rifles, so do certain scientists manage to adapt comfortably to the paradigms of several disciplines. How do they do it? When questioned, these generalists always express an inner faith in the unity of science. They, too, carry a single paradigm, but it is one taken from a much higher vantage point, one from which the paradigms of the different disciplines are seen to be very much alike, though often obscured by special language.

Kenneth Boulding once remarked that the generalist is like the tourist who, when he sees Bangkok, is reminded of Pittsburgh, because both are cities and have streets with people in them. Like the tourist, he relieves his fear of strange places—strange paradigm systems—by moving to higher and higher levels of generality until all things follow the same familiar, comfortable order. In ascending like this, general systems thinking is following the same human tendency that produces the paradigms in the first place.

The most dangerous pitfall in developing category systems is imagining that one system of paradigms is more "real" than another. For example, stars in the heavens present the same "objective" picture over wide portions of the earth, and every human culture seems to have developed ways of looking at those stars as familiar objects—be they as animals, people, or kitchen utensils. Although each of those systems is different, each culture "really sees" its own pictures, often worships them, and is patently unable to "see" the pictures of another culture. Our astronomers also claim to have a truthful ordering of the heavens, but how can we evaluate the merits of their claims relative to those of the others? If we appeal to "usefulness," then certainly each culture has as much claim to "truth" as any other, for none find the other systems understandable, let alone useful. But if we appeal to some intrinsic "truth," then we are making a religious argument, and how shall we adjudicate the claims of the diverse relgions?

Psychologically, it may be essential for a scientist to have faith in the truth of his own discipline, but such conviction can only diminish his chances of making a revolution or moving to another discipline. For the missionizing interdisciplinarian, such faith is doubly essential. Yet just as the gun impedes communication with the natives, the single model of the interdisciplinarian prevents him from learning about the field in which he uses it.

To be a good generalist, one should not have faith in anything. Faith, as Bertrand Russell once pointed out, is the belief in something for which there is no evidence. Every article of faith is a restriction on the free movement of thought, and thus on the free movement of the generalist among the disciplines. As Reichenbach[8] observed:

> The power of reason must be sought not in the rules that reason dictates to our imagination, but in the ability to free ourselves from any kind of rules to which we have been conditioned through experience and tradition.

The Main Article of General Systems Faith

My advice to any young man at the beginning of his career is to try to look for the mere outlines of big things with his fresh, untrained, and unprejudiced mind.[9]

H. Selye

But nobody can exist without faith in something. Without faith, we could not move one foot in front of the other, not knowing whether the next piece of ground would support our weight. Moreover, we could not even stand still without faith in the continuity of the ground now beneath our feet. The general systems approach does not free us from the need for faith, but only attempts to supplement one set of beliefs with another, in the hope of sometimes being more useful.

On what basis does general systems thinking promise to be useful? The principal answer seems to lie in what Boulding calls "The Main Article of General Systems Faith":

> This is that the order of the empirical world itself has an order which might be called order of the second degree.[10]

About the generalist, Boulding says:

> If he delights to find a law he is ecstatic when he finds a law about laws. If laws in his eyes are good, laws about laws are delicious and are most praiseworthy objects of search.

This faith, this hunger, could be in vain. But if an order of the second degree *does* exist, then surely it will be useful to those who seek order of the first.

In a certain sense, order of the first degree underlies the order of the second, and the primary way of discovering general systems laws is by *induction.* The general systems researcher starts with the laws of different disciplines, searches for similarities among them, and then announces to the world a new "law about laws." The general laws of the disciplines are thus only particular cases to him.

The power of generalization through *induction* is that we can then use the general laws to draw conclusions about cases not yet observed. This is the source of the generalist's power to move from discipline to discipline. Each time he is successful, he provides one more piece of evidence for his belief in order of the second degree.

The main article of general systems faith is, therefore, not entirely based on faith. Faith is needed, however, because this leaping from discipline to discipline does not always work. Why not? Because induction does not always work. Even though it looks as if he is operating on the most general plane, the generalist is, like any scientist, only applying the results of induction. Philosophers have tried for a long time to demonstrate that induction must work, but now the smart ones have given up. As Reichenbach[11] said:

> We need (induction) if we want to establish a general truth, which includes a reference to unobserved things, and because we need it, we are willing to take the risk of error.

But why can't we be more careful? Why not wait until more evidence is in? Why such a hurry? The answer lies in the explosive growth of knowledge, and in the limits the Square Law of Computation places on our brain:

> Even the most renaissance of renaissance men in these days cannot hope to know more than a very small fraction of what is known by somebody. The general systems man, therefore, is constantly taking leaps in the dark, constantly jumping to conclusions on insufficient evidence, constantly, in fact, making a fool of himself. Indeed, the willingness to make a fool of oneself should almost be a requirement for admission to the Society for General Systems Research, for this willingness is almost a prerequisite to rapid learning.[12]

To be a successful generalist, then, we must approach complex systems with a certain naive simplicity. We must be as little children, for we have much evidence that children learn most of their more complex ideas in just this manner, first forming a general impression of the whole and only then passing down to more particular discriminations. Piaget cites his observations that

> `...a child of 4 who did not know his letters and could not read music managed to recognize the different songs in a book from one day or one month

to another, simply by their titles and from the look of the pages. For him, the general effect of each page constituted a special scheme, whereas to us, who perceive each word or even each letter analytically, all the pages of a book are exactly alike.[13]

In verbal matters, adults may lose their ability to grasp wholes before examining parts, masking it with superior analytical ability at reading or listening. Nevertheless, we all retain some ability to function without verbal clues. We can recognize a familiar city block even when it has no signs, and we can sense when something is different even when we cannot specify the change.

To be sure, by foregoing analysis we are exposed to certain errors. We are often mistaken in our impression that we have been on this block before, as further analysis might clearly show. But in science, as Selye also said, "There is a great deal of difference between a sterile theory and a wrong theory."[14] When lost in a slightly familiar neighborhood, we need general impressions as quick guides to more familiar territory. If we are mistaken and find outselves on the wrong block, the mistake may be readily corrected. If we insist on reading every house number on every block, however, we may miss dinner.

No approach, be it analytic or synthetic, can guarantee a flawless search for understanding. Each approach has its characteristic errors. By taking the grand leap based on our faith in order of the second degree, we may often be completely wrong, but at least we shall find out soon enough. If time is of the essence, the slow-but-sure method of analysis may only guarantee that we cannot possibly arrive on schedule. Lord Rayleigh once remarked that:

It happens not infrequently that results in the form of "laws" are put forward as novelties on the basis of elaborate experiments, which might have been predicted *a priori* after a few minutes consideration.

This is the characteristic error of analysis. Though in the long run it always rewards our patience, in the long run, as Keynes noted, we shall all be dead. Therefore, those who are impatient with precise methods are attracted to the general systems approach, but mere impatience is not enough. To be a successful generalist, one must study the art of ignoring data and of seeing only the "mere outlines" of things.

Perhaps the nature of this prescription will be better understood if we look at its very antithesis—the Austrian school inspector described by Wertheimer[15]:

The events are said to have happened in a small Moravian village in the time of the old Austrian empire. An inspector from the Ministry of Education arrived one day to visit the school room. It was part of his duty to make such periodic inspections of the schools. At the end of the hour, after he had observed the class, he stood up and said: "I am glad to see that you children are doing well in your studies. You are a good class; I am satisfied with your progress. Therefore, before I go, there is one question I would like to ask: How many hairs does a horse have?" Very quickly one little nine-year-old boy raised his hand, to the astonishment of the teacher and the visitor. The boy stood up, and said: "The horse has 3,571,962 hairs." The inspector wonderingly asked: "And how do you know that this is the right number?" The boy replied: "If you do not believe me, you could count them yourself." The inspector broke into loud laughter, thoroughly enjoying the boy's remark. As the teacher escorted him along the aisle to the door, still laughing heartily, he said: "What an amusing story! I must tell it to my colleagues when I return to Vienna. I can already see how they will take it; they enjoy nothing better than a good joke." And with that he took his leave.

It is a year later, the inspector is back again at the village school for his annual visit. As the teacher was walking along the aisle with him to the door, he stopped and said: "By the way, Mr. Inspector, how did your colleagues like the story of the horse and the number of his hairs?" The inspector slapped the teacher on the back. "Oh, yes," he said. "You know, I was really very anxious to tell this story—and a fine story it was—but, you see, I couldn't. When I got back to Vienna I wasn't able for the life of me to remember the number of hairs."

The Nature of General Systems Laws

It will be objected that this sharpness or clarity involves certain distortions or misrepresentations, depending on over-simplifications. But this is the perennial dilemma of the teacher: the teaching of facts and figures vs. the teaching of truth. To convey a model the teacher must reify and diagram and declare clearly what cannot be seen at all. The student must "learn" things in order to realize subsequently that they are *not* quite the way he learned them. But by that time he will have gotten into the spirit of the matter, and from this he may arrive at some approximation of the truth, an approximation he will continue to revise all his life long.[16]

Karl Menninger

We have so far discussed the role of analogy, category schemes, generalization, and other tools of general systems thought. Now we should like to explain the use of "laws" throughout this book. Before doing so, it will be necessary to remind ourselves of some aspects of scientific laws, aspects that are not always emphasized in standard works.

In particular, we want to be reminded that:

The paradigm of a scientific assertion is "If so . . . then so." [17]

One of the reasons we forget the *conditional* nature of scientific laws is that those laws are often stated in a shorthand manner, with the "If so . . ." part simply dropped or abbreviated. This part *must* be abbreviated because of the enormous length it would have if we seriously attempted to write it all out. For example, one way of stating the First Law of Thermodynamics is:

Total energy in a system is constant.

We could elaborate this statement in operational terms by something like this:

If we have a system to which energy is neither brought nor taken away, and if we make measurements of the total energy of that system, while in the process of measurement neither bringing nor taking away energy, then every measurement will give us the same value.

This statement could be elaborated further, but it is clumsy enough as it stands. Certainly it would be harder to remember than the previous statement, and trying to be even more precise would only make it worse.

Still, we sometimes need very precise statements of the "if so" conditions under which a certain "then so" will hold. For instance, suppose we do measure the energy in a system and find that all measurements do not give the same value. We may then conclude:

1. the Law of Conservation of Energy does not hold for this system;
or
2. some energy was brought in or taken away;
or
3. our measurement was not correct.

Most likely we shall preserve the law, for the law represents a codification of a great many previous experiments. Although, in theory, one negative experiment would force us to *reject* the Law of Conservation of Energy, in practice we shall probably do no such thing.

In the first place, we are most likely to suspect our measurements. In that case, the law may be used as a rule for defining "measurements of the total energy":

If we have a system to which energy is neither brought nor taken away, and if we make measurements of some attribute of that system, without bringing or taking away energy, and if the attribute measured is not constant, then that attribute is *not the total energy of the system.*

Alternatively, we may conclude that some energy was taken away or brought to the system, in which case the law is a partial definition of a "closed system," or a reminder to look for an "opening":

If we make measurements of the total energy of a system and if we find that the total energy changes from measurement to measurement, then the system is not closed.

A more drastic approach would be to change the definition of "total energy" so as to preserve the law. This was actually done to preserve the law when Einstein made his famous assertion about the equivalence of matter and energy:

$$E = mc^2$$

This equation could be interpreted as saying that matter *can be converted into* energy (and vice versa), or that matter *is a form of* energy. The second approach preserves the Law of Conservation of Energy. The first also preserves it, but with an added "if so" clause:

. . . and if no conversions between matter and energy take place within the system . . .

We see, then, how many different roles laws play in scientific thinking. They prescribe guides to measurement, they define the terms within them, they remind us to look for things we have not noticed, and they predict behavior. They also provide a sort of rallying point around which we can discuss ways of measurement, the meaning of terms, and heuristic, or problem-solving technique. The same law can do all of these things, though obviously not at the same time. Learning to think scientifically is not just a matter of remembering the laws, but of knowing when to use which law and in what way.

If a law has many if–so clauses, it will be difficult to remember when to use it, for each if–so clause limits the scope of application. The fewer the if–so clauses, the more *general* the law. When faced with the problem of adding an if–so clause or changing definition of a term, the choice will usually be made toward redefining the term. Thus, while it is claimed that the Law of Conservation of Energy has withstood more than a century of testing, it has only been salvaged by a succession of redefinitions of "energy."

When measurements are found incompatible with a well-established law, the last thing to be changed is the law itself—contrary to the popular impression that one negative case invalidates a scientific law. Indeed, we might well forge a new general systems law that says:

When the facts contradict the law, reject the facts or change the definitions, but never throw away the law.

This could be called *The Law of Conservation of Laws*.

Science obeys the Law of Conservation of Laws because scientific laws contain too much valuable information simply to be jettisoned when they are "invalidated." In the process of preservation, however, laws may become pickled in great lists of conditions, definitions, and exceptions. Eventually, they lose their original flavor as shorthand summaries of inductive knowledge, even as they yield more precise answers to ever-narrower questions.

"General systems laws," as used in this book, are not designed to yield *answers*; therefore, they can afford occasionally to be wrong. Presumably, general systems laws will never be used for precise conclusions without checking the insights they provide. Therefore, rather than make general systems laws more *precise* by pickling them with qualifying conditions, they are made more *memorable* by keeping their original flavor free of complications. Moreover, wherever possible, we try to increase their memorability by clever phrasing or a catchy name. Perhaps we would do better to call them "aphorisms," but, then, "law" is such a catchy name.

For some unknown psychological reason, one of the most memorable ways to phrase a law is in the form of a *prohibition*, a *contradiction*, or even better, a *paradox*. One formulation of the Law of Conservation of Energy was:

It is impossible to build a perpetual motion machine.

When it was discovered that the First Law did not exclude a certain kind of perpetual motion—though the Second Law did—the definition of "perpetual-motion machine" was changed to what we now call a perpetual-motion machine "of the first kind." What that means, of course, is that the kind of perpetual-motion machine we call the "first kind" is the kind the First Law says we cannot build—a neat application of the Law of Conservation of Laws.

Many of our general systems laws will be stated in several forms: as definitions, as measurement guides, as heuristic devices, and particularly in the more memorable negative forms. Often we shall present the law in an approximate form in order to simplify the discussion, and to draw attention to the conditions that will accrue to the more elaborate form without burdening the reader's mind with too many "first kinds," "second kinds," and so forth. A wrong law may be useful, but no law is useful if you don't remember it when you need it. Therefore, our laws should not be taken as constraints to thought, but as *stimulants.*

The memorability of our laws will also be enhanced if we do not fail to give illustrative examples. We hope to avoid the disease of hollow generalization, for it is not just the large generalization, but "the large generalization limited by a happy particularity, which is the fruitful conception."[18] For each law, we shall endeavor to give not just one, but two "happy particularities," sometimes as research problems at the end of the chapter. Anything that aspires to the highfalutin' title of "general systems law" should be demonstrably applicable to at least two concrete cases—the one from which it came and one more for security.

General systems literature has not always conformed to this principle. Perhaps, then, we should elevate it to a general systems law, which we might call the *Law of Happy Particularities*:

Any general law must have at least two specific applications

or, as Elise Boulding tells her husband whenever he flies too far from facts:

If you are going to be the great integrator you ought to know something.[19]

Out of courtesy to my colleagues, I shall refrain from citing two specific applications of this law. In doing so, I am providing one example myself. The reader will surely find others as he proceeds.

Overgeneralizing is the error of a fool or a hero, depending on your point of view. But just as an excess of courage leads to overgeneralizing, so does an excess of caution lead to undergeneralizing. Balancing the Law of Happy Particularities is the *Law of Unhappy Peculiarities*:

Any general law is bound to have at least two exceptions

or, to put the emphasis in a negative way:

If you never say anything wrong, you never say anything.

The reader is invited to look for the two exceptions to the Law of Unhappy Peculiarities.

While the Particularities and Peculiarities Laws are applicable to any generalizing behavior, there are also laws applying to the typical "systems" part of general systems thinking. Again, there are two complementary errors—composition and decomposition. An example of a composition error is this:

I stand on a bridge and spit in the river. Seeing that it makes no noticeable difference in the purity of the water, I go to the polls and vote against the municipal bonds for a new water-treatment plant.

An example of a decompositional error, on the other hand, would be this:

I stand on the bridge and notice that the river is clean, so I conclude that nobody spits in it.

To warn us from these two errors, we have two laws. The *Composition Law*, which goes back at least to Aristotle, says:

The whole is more than the sum of its parts.

The *Decomposition Law*, on the other hand, says:

The part is more than a fraction of the whole.

Notice that the two laws seem contradictory, which should made them hard to forget.

Why should we want to remember them? Of what use, really, are general systems laws? Because they are so general, and because systems are so complex, we shall not find them very helpful at making exact predictions. But because they *are* so general, and because systems *are* so complex, general systems laws can help us avoid the grand fallacy on the way to an exact prediction. "It isn't what we don't know that gives us trouble, it's what we know that ain't so."

Varieties of Systems Thinking

The main role of models is not so much to explain and to predict—though ultimately these are the main functions of science—as to polarize thinking and to pose sharp questions. Above all, they are fun to invent and to play with, and they have a peculiar life of their own. The "survival of the fittest" applies to models even more than it does to living creatures. They should not, however, be allowed to multiply indiscriminately without real necessity or real purpose.[20]

In his remarks on mathematical models, Kac provides an outline of the joys, uses, and applications that could correctly be applied to the models of general systems. He implies that there are three sorts of activities involving models:

1. Improving thought processes—"to polarize thinking and to pose sharp questions"
2. Studying special systems—"real necessity or real purpose"
3. Creating new laws and refining old—"to invent and play with."

We can use this framework to review the "general systems approach" as outlined rather loosely in this chapter, and as a preface for the remainder of this work. We can start with *improving thought processes,* for this is the contribution that will be most used by most people. We are not all engaged in studying special systems, and even fewer of us are engaged in creating new general systems laws, but most of us are engaged in thinking.

The contribution of the general systems approach to thinking is perhaps best seen in the way a generalist approaches some new subject. Students should be particularly interested in this application of general systems, for they have new subjects thrust upon them each semester. Unhappily, the trauma from four years of new subjects often paralyzes the mind, and many graduates remove their cap and gown uttering an oath never again to learn a new subject. The general systems approach promises to make learning new subjects less traumatic, so that education may begin to give a taste for, rather than a disgust with, learning.

How does the generalist approach a new subject? Suppose, for instance, that he decides he must learn something about *economics.* He might find a textbook on the subject by seeing what is used in the introductory course in the local university, or he might simply browse among the economics books in his local library. When he opens such a book, however, he is not starting from scratch. He knows many general paradigms for thought and communication, and he is clever enough to penetrate their economics disguise.

For example, if he happened to pick up Samuelson's *Economics,*[21] he would find in Chapter 2 a variety of "production-possibility" curves, such as we see in Figure 2.2. Almost without explanation, he recognizes these as special cases of a more general representation, the *state space* (which we shall discuss in a few chapters). What the economist calls "the production-possibility frontier," the generalist recognizes as a *family of systems* all possessing a certain property, and any point on that curve representing a particular system within that family. He recognizes that the movement from one point to another in this state space is what he calls a *line of behavior,* so that all he knows about such lines may immediately be transferred to this new situation.

Because of the nature of general systems laws, the content that is transferred may actually be quite small, from the economist's point of view. Nevertheless, the generalist is already at an advantage over his colleagues because, like the tourist in Bangkok, he is not afraid of the unfamiliar. He has named the beast, and he has thus begun to tame it.

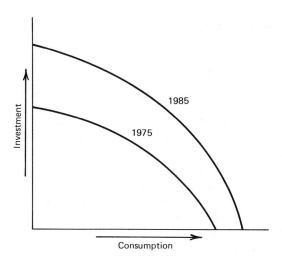

Figure 2.2. The economist's "production possibilities" are the generalist's "state space."

The generalist, then, has certain categories of thought that, because of their general nature, are not going to fail him completely in the study of any new field. He has special words in his vocabulary, words such as stability, behavior, state space, structure, regulation, noise, and adaptation, which he can relate to the words of the specialist. If he is wise, he will refrain from saying, "Oh, that's nothing but a line of behavior in a two-dimensional state space." Instead, he will make the translation internally and then surprise the specialist with the "sharp questions" he is able to ask.

When the generalist encounters laws in the special field, he will often be able to relate them to the general systems "laws" he knows. He identifies the special assumptions that have made his general systems laws into laws of economics, or whatever. For instance, he will immediately recognize the economist's *Law of Diminishing Returns* as a special instance of the *Law of Limiting Factors*. Of course, he does not boast that the economic law is "only a special case" of the general law, especially since he realizes that the general law was probably dependent on the economic law for its discovery. While *for him* the one law is a special case of the other, *in genesis* the general systems law was probably induced from the economic.

The general systems approach, then, can engender a parsimony of thought for the study of subjects. A similar economy is introduced in the study of situations, or *special systems.* In our experience, the general systems approach has provided a starting point for the study of a myriad of *information systems,*[22] *complex machines,*[23] *social systems,*[24, 25] *individuals and work groups,*[26] and *systems for education.*[27] Others have found the general systems approach useful in meteorology, political science, biology, sociology, psychiatry, ecology, engineering, and, in fact, just about any discipline you can name. The reader who is interested in specific examples will find a gold mine in the collected yearbooks of the Society for General Systems Research.[28] He should be warned, however, that these collected articles represent but a small fraction of the applications of general systems thinking to special systems situations, because most of the time the thought is applied and not written up for publication in some journal. No, most of the application of general systems thinking is not done by professional academics, but by ordinary people going about their daily business.

The *General Systems Yearbook* also contains examples of the third kind of general systems activity—*creating new laws and refining old.* This activity we call *general systems research,* as opposed to *general systems thinking* and *general systems application.* Of the three activities, general systems research is practiced by the smallest number of people, and therefore is really the interest of specialists. We cannot say much more about how general systems research is conducted than we can say about how research is conducted in other fields. We do have some *very general systems laws,* such as the Law of Happy Particularities, which are really laws about how to conduct general systems research. Mostly, however, *research* in general systems is carried out in the same mysterious ways as research in any discipline.

The general systems movement did not start out as a discipline but is probably ossifying into one. In the foreward to his 1969 book,[29] von Bertalanffy surveyed three decades of general systems activity by warning that we may find

   . . . systems theory—originally intended to overcome current overspecialization—another of the hundreds of academic specialties. Moreover, systems science, centered in computer technology, cybernetics, automation, and systems engineering, appears to make the systems idea another—and indeed the ultimate—technique to shape man and society ever more into "megamachine" . . .

Years ago, in the innocent days before the move toward academic bureaucratization and military funding had begun to penetrate the Society for General Systems Research, I received a letter addressed to "The Society for *Gentle* Systems Research." As I watch in horror what is becoming of the "systems movement," I often recall that letter, and wonder whether there will still be room for the gentle people, and whether we shall perhaps help to build not "megamachines," but gentle systems.

How will it come out? Most likely, it will come out the way all movements come out, by killing its prophets and reversing their words. It has already reached the point of no return, but like any fanatic, I cannot resist making one last try. This work, which tries to bring general systems back to the ordinary people for whom it was conceived, is that desperate gesture.

QUESTIONS FOR FURTHER RESEARCH

1. *Anthropology*

In Mexico, there is a village of blind people, which was studied by a blind anthropologist. Speculate on the cultural categories that this village might have different from a similarly situated village of sighted people. Would the blind anthropologist find that his blindness gave him more in common with the villagers than his being Mexican would have? What would a one-eyed anthropologist have found different if he had studied this village?

2. *History of Science*

A summary of organismic analogies throughout history is given in:

Howard Becker and Harry Elmer Barnes, *Social Thought From Lore to Science,* 2nd ed. Washington, D.C.: Harren Press, 1962.

Has organismic thought ever been helpful to the advance of science? How would you account for the violence with which vitalists and mechanists are always at each other's throats? Are there similar controversies going on today that you can relate?

3. *Molecular Biology*

One of the more recent scientific revolutions has been in molecular biology. In that field, one of the great landmarks (for which its authors received the Nobel Prize) was the double-helix model of DNA. We are fortunate in having the highly personal account given by one of them:

James D. Watson, *The Double Helix.* New York: Atheneum, 1968

plus a number of other accounts issued, as it were, in rebuttal to what some have called a most unflattering account of the scientist at work. Using examples from this book and the controversy surrounding it, discuss the changing paradigms in molecular biology, and what they meant in practical terms to these workers.

4. *Order of the Second Degree*

"Order of the second degre" is a slippery concept. We are never sure of the source of the order we see, and a nice example of the difficulties we encounter in seeking order in order is the observation that first digits in all sorts of tables tend not to be evenly distributed but overly laden with the smaller digits, especially the ones. Ralph A. Raimi has surveyed this problem in a *Scientific American* article entitled "The Peculiar Distribution of First Digits" (December 1969, **221,** 15) Study this problem and give your opinion as to the source of this order of the second degree.

5. *Laws about Laws*

As the inventory of human thought accumulates vast piles of ideas, people make attempts to manage the collection. One form of attempt is to catalog the great ideas, and record their development, as in

> Philip P. Wiener, *Dictionary of the History of Ideas.* New York: Charles Scribner, 1973.

Using the theories and ideas presented in this volume as the source of your data, derive some general systems laws—which are, after all, another approach to managing the same inventory.

6. *Order of the Third Degree*

We now have two pairs of general systems laws dealing with errors. Examine the structure of these laws and see if you can make an appropriate leap of faith.

7. *Environmental Pharmacology*

Patients are often given several drugs at the same time, and sometimes the combination results in an undesirable effect because one drug inhibits or stimulates the metabolism of the other.

Sometimes the "drugs" are given unintentionally, to people who are not "patients." As the number of different chemicals appearing in our environment increases, what sorts of effects are likely to "emerge," to what extent can these emergences be predicted, and to what extent will they be predicted by the people putting the chemicals in the environment?

> *Reference*: A. H. Conney and J. J. Burns, "Metabolic Interactions Among Environmental Chemicals and Drugs." *Science,* **178,** 576 (November 1972).

READINGS

RECOMMENDED

1. Kenneth Boulding, "General Systems as a Point of View." In *Views of General Systems Theory,* Mihajlo D. Mesarovic, Ed. New York: Wiley, 1964.
2. H. G. Wells, "The Country of the Blind." In *The Country of the Blind and Other Stories.* New York: Nelson, 1913 (also reprinted in several collections, such as *The Complete Short Stories of H.G. Wells).* (First edition, 20th impression) London: Bern, 1966.

SUGGESTED

1. Thomas Kuhn, *The Structure of Scientific Revolutions.* Chicago: University of Chicago Press, 1962.
2. James D. Watson, *The Double Helix.* New York: Atheneum, 1968.

3

System and Illusion

The real world gives the subset of what *is*; the product space represents the uncertainty of the *observer*. The product space may therefore change if the observer changes; and two observers may legitimately use different product spaces within which to record the same subset of actual events in some actual thing. The "constraint" is thus a *relation* between observer and thing; the properties of any particular constraint will depend on both the real thing and on *the observer*. It follows that a substantial part of the theory of organization will be concerned with *properties that are not intrinsic to the thing but are relational between observer and thing.*

W. Ross Ashby[1]

A sweet disorder in the dress
Kindles in clothes a wantonness:
A lawn about the shoulders thrown
Into a fine distraction,
An erring lace, which here and there
Enthralls the crimson stomacher,
A cuff neglectful, and thereby
Ribbands to flow confusedly,
A winning wave (deserving note)
In the tempestuous petticoat,
A careless shoe-string, in whose tie
I see a wild civility,
Do more bewitch me, than when art
Is too precise in every part.[2]

Robert Herrick

A System Is a Way of Looking at the World

The understanding of the symbol did not necessarily pre-suppose an understanding of its conventional application. This is why I put up such a strong resistance when grandmama wanted to teach me the notes of the scale. Using a knitting needle, she pointed to the notes on the stave; this line, she tried to

51

explain, corresponded to that note on the pianoforte. But why? How could it possibly do that? I could see nothing in common between the ruled manuscript paper and the keys of the instrument. Whenever people tried to impose on me such unjustified complusions and assumptions, I rebelled; in the same way, I refused to accept truths which did not have an absolute basis. I would yield only to necessity; I felt that human decisions were dictated more or less by caprice, and they did not carry enough weight to justify my compliance. For days I persisted in my refusal to accept such aribitrary regulations. But I finally gave in: I could finally play the scale; but I felt I was learning the rules of a game, not acquiring knowledge. On the other hand I felt no compunction about embracing the rules of arithmetic, because I believed in the absolute reality of numbers.[3]

Simone de Beauvoir

What is a system? As any poet knows, *a system is a way of looking at the world.*

The system is a point of view—natural for a poet, yet terrifying for a scientist! As soon as he recognizes the path we are about to take, he rebels, like Simone de Beauvoir, as if we are about to impose some falsehood on him. To speak of systems in this way is to play a game, not to acquire knowledge. Knowledge is "truth," knowledge is "reality." If two scientists viewing the same scene have different "systems," then science will be "no better than" poetry, where one man can see a "wild civility" in another man's "sloppy clothes."

Very well, let us assault fears. Look at Figure 3.1. What do you see? A young maiden, "come forth, like the spring-time, fresh and green . . ."? A crinkled hag, whose rose-buds are gathered and whose youth and blood are spent? No matter. Whichever you see, look again until you see the other, if you see neither, all the better for my argument.

I have used this figure for over a decade as a demonstration of the power of "point of view." Year after year, some see youth; some, age; a few, nothing. Yet it is not *what* we see that is important in this demonstration, but *how we feel* about what we see. After each class, students come to my office to coax me into admitting that I know it is *really* an old (young) woman and that those who see a young (old) woman were fooled. But *they* are the fools—for believing that other people's points of view are foolish, or less *true* than their own.

Some students are more humble. They come merely to *find out* which is the true picture. They have seen the error of egocentrism and are operating on a higher plane. Their foolishness is thus all the greater, for they fail to see that the concept of observer–independent

truth is the ultimate egocentrism. If there were an independent truth, who are *they* to know it?

Egocentrism is a form of animism, and animism, in turn, of vitalism. Through centuries of painstaking effort, scientists have worked to get rid of thoughts like

If I were a planet, sailing through space, how would I be attracted to the great mass of the sun?

If they cannot help thinking that way, at least they have learned to keep it to themselves. Biologists face the same problem, but it is more painful to them, for they feel closer to their subjects:

If I were an oyster, would I be irritated by a grain of sand?

and closer

If I were a frog, would I be frightened by a shadow?

and closer

> If I were a dog, would I like a pound of hamburger?

Psychologists, of course, have it even worse, but the ultimate difficulty, which we all share, is to get rid of such thoughts as

> If I were nature, would I tell lies?

or

> If I were nature, would I throw dice?

How would we known how nature (read "reality") feels? Or if there is any more meaning in speaking of how nature feels than in empathizing with a planet, an oyster, a frog, or a dog?

Each such animism has barred the way to scientific progress, yet each would have been far easier to eliminate had it been totally without use. We *can* get insight into the ideas of force and motion from our internal responses to situations. Newton could, and we still teach physics that way. We can get insights into biological laws from such subjective experiences as irritation, fright, and liking. And we can make progress in science by believing in the reality of the external world.

At this point, the "realists" in such a discussion will quote Einstein:

> The belief in an external world independent of the percipient subject is the foundation of all science.[4]

This quotation might have been used instead of Figure 3.1, for we each read it according to our preconceived notions. Note well that Einstein did *not* say

> An external world independent of the percipient subject is the foundation of all science.

Einstein was a careful man, a careful *scientist.* He did not say that an external world is essential, but that "belief in" an external world is essential. And he was right. Yet it was Einstein who put forth the Theory of Relativity, which rocked the scientific world just *because* it was based on the premise that we could only know the external world through our perceptions.

"Belief in an external world independent of the percipient subject" is a heuristic device, a mental tool to aid in discovery. Like all heuristic devices, it cannot tell us when and where it can be successfully applied. We learn a little rhyme—"*I* before *E*, except after *C*"—that helps us learn to spell, but fails us on "either and neither. . . ." Or, as the little boy said, "Today we learned how to spell 'banana,' but we didn't learn when to stop."

We have encountered this idea before. Mechanics alone cannot say which systems will yield to mechanical analysis. Mathematics cannot tell us its range of successful application. In honor of that little boy, we can elevate his idea to a principle, The Banana Principle:

Heuristic devices don't tell you when to stop.

We have a scale of ascending values for heuristic devices, depending on how far you go before you must stop. Going from the narrow to the broad, we find: "idea," "concept," "rule," "principle," "law," "reality," "truth." The further along this scale, the less we notice that a heuristic device is a device. We forget the Banana Principle and think that we can go on using the device forever, and the more success we have, the more sure we become.

But the more sure we are, the more likely we are to suffer an illusion, for the illusion consists

. . . in the conviction that there is only one way of interpreting the visual pattern in front of us. . . . The most famous story of illusion in classical antiquity illustrates the point to perfection; it is the anecdote from Pliny, how Parrhasios trumped Zeuxis, who had painted grapes so deceptively that birds came to peck at them. He invited his rivals to his studio to show them his own work, and when Zeuxis eagerly tried to lift the curtain from the panel, he found it was not real but painted. . . .[5]

Our aim is to improve thinking. "The belief in an external world independent of the percipient subject" is one of the most powerful thinking tools we have. We have not the slightest intention of discarding such a powerful tool, any more than we intend to discard analogical thinking. Nor could we discard it if we wished to. We shall soon enough lapse into speaking in the more familiar terms of an independent reality—for a while we should like to dwell on the complementary tool: relational thinking.

There may actually be "real objects" out there in the world, but if there are, it is not because we perceive them as real. Perception responds just as well to illusion as to reality, and many of these perceptions are so deep that we are essentially powerless to unlearn them, even in illusion situations.[6] Similarly, there may be "real laws of nature," but if there are, our very strong belief in their existence *may be preventing their discovery.* Therefore, let us see what we can learn if we occasionally suspend belief in "independent reality," never forgetting, of course, that this, too, is but a heuristic device.

Absolute and Relative Thinking

There is, in this regard, a pertinent story about the great American linguist and anthropologist Edward Sapir, who had allegedly been working with an informant on an American Indian language with a grammar that he was having trouble sorting out. Finally, he felt he had caught on to the principles involved, and to test his hypothesis he began making up sentences in the language himself. "Can you say this?" he would ask his informant and would then produce his utterance in the informant's language. He repeated this several times, each time composing a different expression. Each time his informant nodded his head and said "Yes, you can say that." This apparently was confirmation that he was on the right track. Then an awful suspicion crossed Sapir's mind. Once more he asked "Can you say this?" and once more received the answer "Yes." Then he asked, "What does it mean?" "Not a darn thing!" came the reply.[7]

It is possible to speak or write perfectly *acceptable* things that do not mean anything. If we study some meaningless statements, we shall better understand how to speak meaningfully, for the exception does not *prove* the rule, it teaches it.

"The exception proves the rule" is an excellent starting point for a discussion of meaningless utterances. "Proof" in its original sense was

a test applied to substances to determine if they are of satisfactory quality.

We retain this meaning in the "proofs" of printing or photography, in the "proof" of a whiskey, and in "the proof of the pudding." Over the centuries, the meaning of the word "prove" began to shift, eliminating the negative possibilities to take on an additional sense:

To establish, to demonstrate, or to show truth or genuineness.

Although the meaning of the critical word was changing, the saying was preserved. As a consequence, we have to contend with illiterates who delight in parroting

The exception *proves* the rule

whenever we contradict one of their favorite prejudices.

Statements in a language have meaning only *in relation to* certain accepted meanings of the words in them. "Accepted meanings" implies *somebody* doing the accepting—an observer. If I say

The exception proves the rule

in front of a large class, there will be a division in understanding, just as there was a division of perception over Figure 3.1. Some will believe I have uttered nonsense, while others will understand:

The exception puts the rule to the test.

The appearance of *absolute* meaning in certain statements comes because there is *almost universal* agreement on the meanings it contains. Consider, for example, the following passage:

General Motors exists to put out cars, not metal scraps, although it extrudes both. Universities exist to produce educated persons and scholars, not retired professors or academic failures.[8]

That seems incontrovertible, but now consider what we would have thought had Miller written:

Beavers exist to control floods, not to produce piles of wood chips. The oceans exist to produce fresh fish, not mud deposits or dead whales washed ashore.

With "man-made" systems, we talk about "purpose," whereas such language is forbidden for "natural" systems. Yet much of the dissatisfaction with our man-made systems stems precisely from disagreement about what the "purpose" of the system is: that is, what the system "really" is. The answer, of course, is that the system has no "purpose," for *"purpose" is a relation,* not a thing to "have." To the junk dealers, General Motors *does* exist to put out scrap metal, yet the stockholders probably couldn't care less whether General Motors is producing cars or string beans, as long as it is producing profits.

Or consider the university. There has been much talk about university reform, but we have seen little action. Why not? At least part of the reason is the failure to recognize that a university is many things to many people. Certainly one of the *most important* social functions of the American university *is* to turn out academic failures, people who will be resigned to taking less lucrative and less prestigious roles in the class structure. As I can attest from inside observation, some professors know that the *real* reason for our institution is precisely to provide us with graceful retirement—even on the job!

What Miller is talking about, then, is not *the* reason for existence of these institutions, but a more or less official *public* reason, much like the public agreement on the meaning of a word. Miller knows this, and need not qualify every statement to death, as in:

From the point of view of most people, most of the time when they think about General Motors they think in terms of producing cars, even though some people, some of the time, have different points of view about the purposes of General Motors.

It is much more forceful to speak in absolute terms, as if there were one single real true "purpose" of General Motors. Most of the time, absolute speech will not get us into trouble, though we may learn something if we bother to examine the relative nature of some seemingly absolute statements.

A simple example of absolute thinking is seen in answers to the question:

What happens to the reading on a candy thermometer if we suddenly plunge it into hot water?

Everyone knows the reading will rise, but if you try this experiment and observe very closely, you will see that before the rise, the reading actually *drops* for a moment. Few people have ever observed this drop, not because it is difficult to see, but because they are not looking for it, because they *know* the reading will rise.

They may know it will rise because they have been told so, in which case they will more easily accept being told otherwise. But those who know the "theory" behind the thermometer will be harder to convince. They know "better"—which is to say that their illusion is stronger. They will argue that the reading must rise because

... the reading measures the expansion of the mercury, and the mercury expands when heated.

There are at least two concealed absolutisms in this simple statement. One has to do with the *time scale* of the observation. The statement seems to imply instantaneous expansion of the mercury, but what it could more precisely say is

... and the mercury expands as its temperature rises, which takes a relatively short time on the human time scale.

We know, of course, that it does take time for the mercury to warm, otherwise we would not hold a fever thermometer in our body for several minutes when we take our temperature.

The time scale explains why the rise may follow a curve such as shown in Figure 3.2, but why does it fall first? The answer lies in the second concealed absolutism, "the expansion of the mercury." The reading does not measure the expansion of the mercury, but the *difference* in expansion between the mercury and the glass. That is, it measures the *relative* expansion of the mercury, not the *absolute* expansion.

When we plunge the thermometer into hot water, the glass, being on the outside, warms first, and therefore begins to expand first. Since the

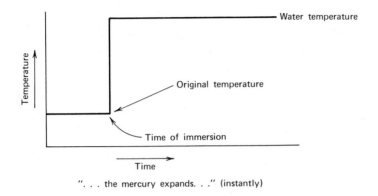

". . . the mercury expands. . ." (instantly)

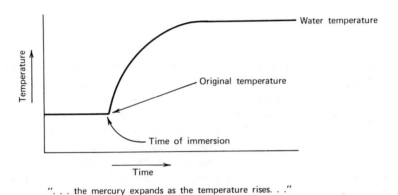

". . . the mercury expands as the temperature rises. . ."

Figure 3.2. Two models of a rising mercury column.

mercury has not yet begun to warm, the mercury has not expanded and thus begins to fall in the tube (not in a fever thermometer, of course, which prevents the mercury from falling back in the bulb, so as to hold the highest reading; you should not plunge a fever thermometer into hot water in any case). The resulting behavior is something like that in Figure 3.3, which indeed shows a drop before the rise.

A thermometer, like our language, is an instrument for understanding our world. When we use the thermometer for simple things, we can use simple language to describe what it does. We do not care how the thermometer "really" behaves—we are satisfied if it behaves the way our simple language says. Glancing at the thermometer mounted outside our window, we say "the temperature is 17 degrees."

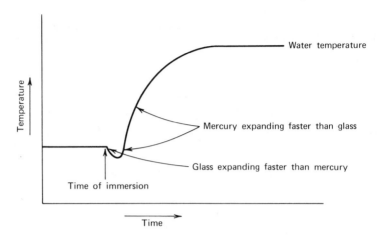

Figure 3.3. ". . . the difference in expansion between mercury and glass . . .

We are not concerned with time-scale effects, even though the temperature might be just changing.

Taking our body temperature, however, we become aware of the time scale because we must use the instrument differently. Finally, when we use a thermometer as part of an automatic-control system, we may need to refine our view to the level of Figure 3.3 and beyond.

One more example: Systems writers sometimes speak of "emergent" properties of a system, properties that did not exist in the parts but that are found in the whole. Other writers attack this idea, saying that emergent properties are but another name for vital essence. Moreover, they can support their arguments with specific examples of "emergent" properties that turned out to be perfectly predictable. Which is right?

Both are right, but both are in trouble because they speak in absolute terms, as if the "emergence" were "stuff" in the system, rather than a relationship between system and observer. Properties "emerge" for a particular observer when he could not or did not predict their appearance. We can always find cases in which a property will be "emergent" to one observer and "predictable" to another.

Demonstrations that a property *could have been* predicted have nothing to do with "emergence." By recognizing emergence as a *relationship* between the observer and what he observes, we understand that properties will "emerge" when we put together more and more complex systems. In other words, the property of "emergence" no longer emerges for us, though it surprises those who take the absolute view. They may demonstrate—*afterwards*—that they *need not* have been surprised: small consolation if the emergent property was an explosion.

How can we avoid fallacies of absolute thought? The key, I believe, is always to remember the human origins of our models, words, instruments, and techniques. Absolute thought is a simplification that serves us well at certain times, on a certain scale of observation, and for certain purposes. When we say something, when we think in a certain way, we are usually following conventional patterns, patterns that will work out well if the situation remains conventional, which most of the time it will.

The conventional situation is well characterized by the following story:

A minister was walking by a construction project and saw two men laying bricks. "What are you doing?" he asked the first.

"I'm laying bricks," he answered gruffly.

"And you?" he asked of the other.

"I'm building a cathedral," came the happy reply.

The minister was agreeably impressed with this man's idealism and sense of participation in God's grand plan. He composed a sermon on the subject, and returned the next day to speak to the inspired bricklayer. Only the first man was at work.

"Where's your friend?" asked the minister.

"He got fired."

"How terrible. Why?"

"He thought we were building a cathedral, but we're building a garage."

In the workaday world, it seems, we just be practical, keep our eye on the ball, and not get lost with our head in the clouds. If we are going to build garages, we just follow our orders and lay bricks the way we always do. But if we are building a cathedral, then we must look behind what we conventionally do.

Who decided to build a garage? To build cars? To turn out educated persons? Who decided that particular modes of thought were more correct than others? Are their decisions still appropriate in this new situation? Our ways of the world have been handed down to us, but not on stone tablets:

Systems are thoroughly man-made. . . . When we include a given relation in a system, or omit it, we may do well or ill; but such an inclusion creates no truth, and such omission indicates no falsity. The justification for one's procedure, in this respect, is purely pragmatic; it depends upon the relevance of what is included or omitted to the purposes which the system is designed to satisfy.[9]

Because we are here more concerned with building cathedrals than garages, we take the point of view that a system, any system, is the point of view of one or several observers. Whether our view—or their view—is "good" or "bad" can be judged only according "to the purposes which the system is designed to satisfy."

A System Is a Set

More often than not, the classes of objects encountered in the real physical world do not have precisely defined criteria of membership.[10]

L. A. Zadeh

Although any arbitrary way of looking at the world—"a lawn about the shoulders thrown," "an erring lace," "a cuff neglectful"—can be a system, we could never say anything general about truly arbitrary systems. Indeed, we may make a definition:

Arbitrary systems: Systems about which nothing general can be said, except that "nothing general can be said."

If, then, we are to begin a general systems approach, we must narrow our attention to some nonarbitrary systems, though in such a way as to force attention to the *reasons* for the nonarbitrariness. These reasons are the *source of the order* that makes systems thinking possible at all—and the most general of them are the wellspring of general systems thinking.

Nonarbitrariness can come from one of two sources. It could be "out there" in the "real physical world," or it could be in the observer. For the present, we shall focus on the observer. We note immediately that "a lawn about the shoulders thrown," "an erring lace," and "a cuff neglectful" do not form an arbitrary system, for they are seen to belong together by the mind of at least one observer—Herrick. Arbitrary systems, in fact, are hard to find, for as soon as we think of one, it becomes somewhat nonarbitrary.

This argument may sound of the utmost impracticality, but please note that it was exactly in this way that Freud started on the road to discovery in psychoanalysis. As a matter of fact, *nobody* has ever demonstrated that he can choose things arbitrarily. Therefore, if we cannot keep structure out of a *conscious* attempt to make *arbitrary* choices, we may find that unwanted structure is creeping into other systems by way of the observer.

The role of observer is usually ignored in systems writing. The most popular way of ignoring the observer is to move right into a *mathematical representation* of a system—a so-called "mathematical system"—without saying anything about how that particular representation was chosen. For example, Hall and Fagen[11] give this definition:

A system is a set of objects together with relationships between the objects and between their attributes.

Where did these objects come from? Hall and Fagen give no clue. They might have dropped from the sky, except *we* happen to know they came from the mind of some observer.

Hall and Fagen rightly emphasize "relationships" as an essential part of the system concept, but fail to give the slightest hint that the system itself is relative to the viewpoint of some observer. The idea of set is very common in mathematics, but contrary to the impression of precision it gives, it is one of the *undefined primitives* in most theories. The mathematics of sets (set theory[12]) tells us much about the properties of sets, but tells us nothing about how observers might choose them.

The *notation* of set theory will be a great convenience to us, if systems are to be sets of things. For instance, Herrick's system could be described mathematically in the following way:

Let *X stand for* a lawn about the shoulders thrown.
Let *Y stand for* an erring lace.
Let *Z stand for* a cuff neglectful.

Then the set in question is denoted by

$$\{X, Y, Z\}$$

which is not very poetic, but that can be an advantage at times.

The primitive act in all conceptual schemes for choosing sets is the simple, finite act of enumeration—we "set" them down, as it were. We may do this by actually producing all the members of the set for inspection, as a chess set or a set of teeth. Usually, however, we specify a set of *names,* which is taken to *represent* some set of "things," which might be other names, as in the Herrick example. Ordinarily, it is easier to display the names than the things, as with the set:

{Statue of Liberty, Eiffel Tower, Lenin's Tomb, Great Wall of China}

We may, however, want to use names to designate set members that we could not possibly display, no matter how much effort we expended. Such a set might be

{The hemlock that killed Socrates, a proof of Fermat's last theorem, a critical mass of uranium}

The first no longer exists, the second has not yet existed, and the third could not exist in one place long enough for us to observe it and remain alive.

Naming nonexistent set members is naturally a source of potential fallacies. Even more pernicious are those set members that do not exist but whose existence we do not question. The anthropologist may speak of "the kinship terminology rules"; the archaeologist used to speak of "Piltdown Man." The muddles created by erroneous or falsified data fully justify the elaborate safeguards erected by science. On the other hand, data for intuitive thinking have not been subject to these safeguards. The most definite impression of a system may be built on the shifting sands of an imaginary set, even when we have actually enumerated each member.

In any event, we rarely enumerate the collections that form the basis for our thinking. Enumeration, which forms the *conceptual* basis for the other operations, has perils of its own, but these are as naught compared with the possibilities for mischief in *derived* methods. Perhaps the nastiest of these methods is the representation of a set by a *typical member*. This method rests on the assumption that the set can be typified, an idea that goes back at least to Plato. Platonists argued that the ideal type is a *better* representation of the set than any enumeration could be, since the actual members of a set could at best be faulty realizations of the ideal type. The ideal type, however, is strictly an observers's mental construction, which may be a useful way to summarize a mass of data. As taxonomists have discovered, it may simply be the primrose path to a *decomposition fallacy*.

Setting down a typical member or members may be troublesome even when they actually exist in the set, because different people may have different ideas about what set they typify. If I write

{Browning, Blake, Byron, . . . }

what do the three dots stand for? Do I mean the set of all English poets whose names start with *B*? Or all English poets? Or all *great* English poets? Or all *great* Englishmen? Or all great poets? You could easily think of a thousand sets I might have had in mind. Indeed, in a literary essay, the ambiguity might be *intentional*. As a basis for scientific work, on the other hand, the method is dangerous, even when the author has a perfectly clear idea and is merely citing typical members as a shorthand. When his own thought is fuzzy, so much the worse.

The three dots in {Browning, Blake, Byron, . . . } imply an "and so on" process, a process that *follows a rule,* a rule that supposedly can be induced without effort from the three exemplars. Rules, either implicit or explicit, form the third method (after enumeration and typical numbers) commonly used to specify sets.

Rules have an advantage over enumeration when the list of members would otherwise be very long. They are superior to typical members when they can be made explicit and operational. Most often, however, explicit rules are used only in *mathematical* operations, such as choosing the set of even numbers. In dealing with the world, rules are often prohibitively difficult to construct.

Computers have a way of exposing flaws in explicit rules. In the attempt to mechanize a classification procedure, we usually discover much more than meets the eye. Cytologists have long been able to select slides of cells with "abnormal" chromosomes; lawyers have always been able to choose precedents "relevant" to a case; and grammarians never had much difficulty classifying sentences according to their "structure." But when they tried to mechanize—to make the rules explicit enough for a computer—cytologists, lawyers, and grammarians discovered that they never knew precisely what they were doing.

Everyone is familiar with classifying sentences according to grammatical structure. In this area, one of the classic computer examples is the sentence:

TIME FLIES LIKE AN ARROW.

Few of us have difficulty recognizing the structure of this sentence: TIME is the subject, FLIES is the verb, LIKE AN ARROW is the predicate. This *seems* to be a purely *grammatical* analysis. It uses only parts of speech, and not the probable *meanings* of the words, which would make it a *semantic* analysis.

When using the computer, things are not so easy. TIME may be a noun, but it may also be an adjective, as in TIME CLOCK. FLIES may be a verb, but it could also be a noun, as in FRUIT FLIES. LIKE may be a preposition, but it may also be a verb, as in I LIKE YOU. Given these possibilities, how do we know that the structure of

TIME FLIES LIKE AN ARROW.

is not the same as the structure of

FRUIT FLIES LIKE A BANANA.

The answer is that we do *not* know. We jumped to a conclusion based on a probable *semantic* interpretation. If the sentence had been

FRUIT FLIES LIKE AN ARROW.

we might have more easily recognized the ambiguity. Instead, this parsing "emerges" from the computer.

Initially, we thought this classification involved only grammatical considerations, but it went much deeper. We were unconscious of the choice process taking place in our own mind, yet even when we feel aware of possible ambiguities, more may be lurking in the shadows. As the computer revealed, there is yet another perfectly *grammatical* interpretation of

TIME FLIES LIKE AN ARROW.

in which TIME is a verb, and the sentence is imperative, analogous to the form of the sentence:

TIME RACES LIKE A TIMEKEEPER

Without the computer to keep us on our toes, we would remain sloppy grammarians, but unaware of our sloppiness.

There remains another subtle difficulty with rules for choosing sets. When we specify a rule of choice, we imply a "choice set": that is, a set of objects to whose members the rule will be applied. Thus, the set of even numbers is *not* the set of *all* numbers divisible by 2 without a remainder, but the set of all *integers* so divisible. In the same way, a rule for selecting cells with abnormal chromosomes starts with the assumption that the observer can recognize the members of the *set of all cells,* normal and abnormal. Choosing this precursor set, however, may be altogether as difficult as dividing it according to the stated rule. Integers are easy enough to recognize, but cells are not so easy. Even integers can be troublesome if they are not displayed explicitly. Consider an equation such as

$$x = 2b$$

Clearly, x must be divisible by 2 without a remainder, but in order to determine whether x is an integer (and therefore even) we must know more about b.

Once again, the grammatical classification of sentences by computer reveals hidden assumptions—in this case, the difficulties of an implied choice set. To select *grammatical* sentences, we must first know how to recognize sentences. To make the choice explicit for a computer, we might say that

A sentence is a body of text that begins with a capital letter and ends with a period.

Applying this rule to the text

The length of the rod was 3.572 meters.

a computer would conclude that

The length of the rod was 3.

is the sentence in question.

One way to resolve the difficulty is by looking ahead to the leftover piece:

572 meters.

We might reject the previous recognition because this piece does not meet our criterion of starting with a capital letter. While this "look-ahead" rule will resolve some cases, it will complicate things, and also leave us unable to deal with such a sentence as

007 spies.

As we proceed with our analysis, heaping one *ad hoc* case upon the other, semantic rules upon grammatical rules upon orthographic rules, we begin to learn what poor little Buttercup knew all along—things are seldom what they seem. Our simplest mental acts are not at all simple. Although not completely rational, neither are they entirely arbitrary. Although we can do them, they are mostly invisible to us. When we do succeed in making ourselves more aware of what is going on inside our own heads, the outside half of general systems thinking will be easy.

Observers and Observations

I have told you these details of asteroid B612 and I have given you its number because of grownups. Grownups love numbers. When you speak to them of a new friend, they never inquire about essentials. They never say to you: "What is the sound of his voice? What are his favorite games? Does he collect butterflies?" They ask you: "How old is he? How many brothers does he have? How much does he weigh? How much does his father make?" Only then do they think they know him.[13]

Antoine de Saint Exupéry

We have, up until now, been intentionally vague about what the set underlying a system was a set *of*. Hall and Fagen, being engineers, made no bones about saying that it was a set of *objects*. Other writers speak of "parts," "elements," "attributes," "components," or "variables." This discord implies that *nobody* knows what a system is a set *of*.

We should not be surprised. All this diversity of names suggests that the members of the system set are one of the *undefined primitives* of systems thinking. Although systems thinkers talk about these members all the time, they never say what they are, any more than the physicist says what "mass" is. In fact, if we do tell what they are, we are no longer talking about systems in general, but about a particular system.

This situation is well characterized by the story about the three baseball umpires. Each was asked in turn how he called balls from strikes.

The first replied "if they cross the plate between the knees and the shoulders, they're strikes, otherwise they're balls."

The second, however, said "If they're balls, I call them balls. If they're strikes, I call them strikes."

"No," said the third. "They ain't nothin' 'til I call 'em."

In deciding on the nature of our primitives, we are the umpire, the sole arbiter. As long as the members of the set "ain't nothin'," our theorizing is strictly contentless—that is, mathematical. As Bertrand Russell remarked, mathematics gets its appearance of truth from not saying what it is talking about.

A mathematical argument cannot be true or false, but, as the mathematician says, only "valid" or "invalid." "Valid," in effect, means that it is *internally consistent*. When we set up a correspondence between the mathematical argument and something "real," then we can speak of that argument as being "true" for that correspondence. Mathematicians generally assume that an invalid argument can *never* be true, no matter what correspondence is made—but that is a philosophical assertion, since it is obviously not a mathematical one.

One of the problems with the mathematical view is that it cannot distinguish between "sterile" and "productive" arguments. A degenerative disease sporadically afflicting the general systems movement is *hypermathematisis:* the generation of grand, sweeping, and valid mathematical theories—often called "general systems theories"—which are as sterile as a castrated mule. They are sterile because they can be applied to *anything* and thus to nothing; but they are doubly sterile because they are indistinguishable—on a mathematical level—from productive theories. They waste our energy and give productive theories a bad name.

How can we prevent *hypermathematisis*? First, we shall follow the admonition of Maxwell:

Mathematicians may flatter themselves that they possess new ideas which mere human language is as yet unable to express. Let them make the effort to express these ideas in appropriate words without the aid of symbols, and if they succeed they will not only lay us laymen under a lasting obligation, but, we venture to say, they will find themselves very much enlightened during the process, and will even be doubtful whether the ideas as expressed in symbols had ever quite found their way out of the equations into their minds.

By using words, we shall sacrifice the appearance of elegance, but we shall stay closer to the things we want to think about.

Second, we shall follow the Law of Happy Particularities and avoid using any mathematical notation unless we intend to use it more than once. Multiple use will permit a little explanation of the mathematical idea and still provide an economy of notation. For instance, we introduce set notation not because

The stature of a science is commonly measured by the degree to which it makes use of mathematics.[14]

By introducing set notation, we do not increase our stature by a single cubit, but merely give ourselves a convenient way to talk about a *delimited range of possibilities*.

Our first happy particularlity for the use of sets is the elaboration of our concept of observer. What an observer does is make observations. These may be sensations on the sense organs of a biological organism, they may be readings taken by instruments, or they might be a combination of the two. An *observation* may be characterized as the act of *choosing an element from a set*, the set of all possible observations of that type for that observer.

In other words, an observer may be characterized by the observations he can make. The notation of sets helps us to recognize that there are two aspects to an observer—the *kinds of observations* he can make and the *range of choices* he can make within each kind. For instance, Herrick might be said to be able to make two kinds of observations, types of dress and types of disorder. His "scope" as an observer may then be characterized by the set

{Dress, Disorder}

His *range* as an observer may be derived from the range, or "resolution level," or "grain," of each part of his scope. Thus, under *Dress* Herrick can distinguish the elements of the set:

{lawn, lace, cuff, ribbands, petticoat, shoestring}

while under *Disorder* he knows at least

{distracted, erring, neglectful, confused, tempestuous, careless, wild, wanton}

In other words, Herrick as an observer could be modeled by a set:

{Dress, Disorder}

which is, in fact, a *set of sets*—"Dress" having six members, and "Disorder" having eight.

Our characterization of an observer may be at once too narrow and too broad. It may be too narrow because we may have excluded some of the scope or failed to make the grain sufficiently fine. We may not be *aware* of the full scope or the full resolution level, or we may not be *interested* in certain possible observations. For example, in psychological experiments, there may be tiny clues that the experimenter does not notice but that the subject does.

In one case, a pigeon was trained to respond to red circles presented on a card in a window. When each card was presented, the apparatus made a slight click, and the click was different for each card. The experimenter thought that the pigeon's scope was

{Color, Shape}

but in actuality it was

{Color, Shape, Click}

The pigeon was responding to the click, and not particularly to the color and shape. The reader who is beginning to be a generalist will note the similarity between the psychologist's view of his pigeon and Miller's view of General Motors.

A *complete* observation by an observer would consist of one selection from each set in his scope. Thus, for Herrick, {lace, erring} would be one complete observation, and so would {cuff, neglectful}. How many such combinations might our idealized Herrick make? Since there are 6 members of Dress and 8 members of Disorder, there will be 6 times 8, or 48 members of

{Dress, Disorder}

including {lace, erring}, {lace, neglectful}, {cuff, erring}, {cuff, neglectful}, and many others.

This set of all possible pairs—this set of sets—is called the "product set," or "Cartesian product," after Descartes, and could be symbolized:

{Dress × Disorder}

It can be read "the Cartesian product of the set Dress and the set Disorder," or "the Cartesian product of Dress and Disorder," or "the product of Dress and Disorder." The full product set is indicated in Figure 3.4, which the reader is invited to finish. We see that although it is a set of sets, it is another example of the use of set notation to *delimit a range of possibilities.*

The reason the product set may be too *broad* a model of our observer is that though he can make each of the component discriminations, he may not be able to make all combinations. We know from the poem that Herrick can recognize {lace, erring}, but perhaps he will be incapable of recognizing an erring shoe string or an erring petticoat. If he cannot, these two pairs, {shoestring, erring}, and {petticoat, erring}, must be excluded from a more precise characterization of Herrick's powers of observation.

In that case, the Cartesian product, Dress × Disorder, is too broad a characterization of Herrick, and if we use it we will be committing an *error of composition.* Using such a model, we might conclude that Herrick could observe things that he is actually incapable of observing— that is, our model might be too general. On the other hand, if we have properly characterized his scope and the grain of each component, then the Cartesian product model will at least *not exclude* any observations he might make. The product set, then, gives us a way of *preventing undergeneralization,* within the assumptions of scope and grain.

We may note in passing that one symptom of general systems *hypermathematisis* is the use of product sets on everything in sight. The Cartesian product converts "all possible discriminations" into "all possible *combinations of discriminations,*" which has great appeal for generalists. If we use Cartesian products willy-nilly, however, we rapidly generate sets of enormous size—which, because the Cartesian product generates all possible combinations, we call "combinatorial" size. General systems theories that fail to take account of the Square Law of Computation may make perfectly general but vacuous laws be-

Dress = {lawn, lace, cuff, ribbands, petticoat, shoestring}
Disorder = {distracted, erring, neglectful, confused,
tempestuous, careless, wild, wanton}
{(lawn, distracted), (lace, distracted), (cuff, distracted),
(ribbands, distracted), (petticoat, distracted), (shoestring,
distracted), (lawn, erring), (lace, erring), (cuff, erring),
⋮
(ribbands, wanton), (petticoat, wanton), (shoestring, wanton)}

Figure 3.4. The Cartesian product, Dress × Disorder.

cause they exceed the computational capacity of any imaginable system.

In our model of an "observer," we shall remind ourselves from time to time how much computational capacity our model requires. Notice, however, that we have no requirement that our "observer" be able to make individual observations (the members of Dress and Disorder) "correctly." Because these are our primitive, undefined elements, the word "correct" applied to them is meaningless. All he must be able to do is to recognize two sensations or measurements as being "the same"—and he is the final arbiter. In other words, "they ain't nothin' 'til he calls 'em."

The Principle of Indifference

"If you call a tail a leg, how many legs has a dog?"
"Five?"
"No, four. Calling a tail a leg doesn't make it a leg."
                                                               —Attributed to Abraham Lincoln

We may not speak of an observation as being correct or incorrect. Without some notion like "correctness," however, we shall find it difficult to say much about observers and their observations. Therefore, we want to introduce a concept of consistency: that is, the compatability of one set of observations with another.

Clearly, as Lincoln pointed out, a notion of consistency cannot depend on how the observer *names* his observations. If Andrew Marvell calls something {ribbands, tempestuous} and Herrick says it is {cuff, careless}, we do not on that account want to say that they are inconsistent observers. Otherwise, we should have to say that {*manchette, négligente*} is inconsistent with {cuff, careless} merely because one is in French and the other in English.

We may state this idea in the form of a *Principle of Indifference:*

Laws should not depend on a particular choice of notation.

The Principle of Indifference is a powerful reasoning instrument. Consider the case of a systems researcher who derived a formula purporting to measure the *difficulty of a selection process*. He had expressed the difficulty in terms of

S = percentage of items selected.
R = percentage of items *not* selected (rejected)

Although the formula was rather long and involved, I was able to eliminate it as implausible in less than 15 seconds, by applying the Principle of Indifference.

The reasoning went like this: Suppose his formula had been

$$D = R^2$$

where D is the difficulty of doing the selection. Of course, the actual formula was much more complicated, but the reasoning is the same. Suppose, for instance, the problem was to separate 10 sheep from 90 goats (100 animals). Then

$$S = \text{percentage of sheep} = 0.1$$
$$R = \text{percentage of goats} = 0.9$$
$$D = R^2 = 0.9^2 = 0.81$$

Now suppose I simply look at the problem the other way and say I am trying to separate 90 goats from 100 sheep. Then

$$S = \text{percentage of goats} = 0.9$$
$$R = \text{percentage of sheep} = 0.1$$
$$D = R^2 = 0.1^2 = 0.01$$

In other words, according to his formula, it was intrinsically easier to separate the goats from the sheep than to separate the sheep from the goats! Yet if that were the case, we could separate the goats from the sheep and then, after they were separated, say "Oh, I changed my mind. I was actually separating the sheep from the goats."

Since his formula for D was supposed to calculate *the best you could possibly do,* this is intuitively a ludicrous conclusion. By contrast, had his formula been

$$D = R^2 + S^2$$

then separating sheep from goats would give a difficulty of

$$D = 0.1^2 + 0.9^2 = 0.82$$

while separating goats from sheep would give

$$D = 0.9^2 + 0.1^2 = 0.82$$

This formula, at least, obeys the Principle of Indifference. It is indifferent to what we pretend we are doing, and gives the same value either way. It, too, could be a *wrong* formula, but not on the basis of the Principle of Indifference alone. With his original formula, though, the Principle of Indifference allowed me to separate the wheat from the chaff, and suggest that he throw away his ridiculous formula.

A rose by any other name *should* smell as sweet, yet nobody can seriously doubt that we are often fooled by the names of things. During and after a revolution things are often renamed just to change thinking patterns. In seventeenth-century England, for example, one religious person renamed himself (or herself) "Put-Thy-Trust-in-Christ-and-Flee-Fornication Williams." In France, in the nineteenth century, "queen bee" was changed to "laying bee," as part of the effort to expunge all records of royalty. In twentieth-century Russia, Tsaritsyn became Stalingrad, and later became "Volgograd," as first the Tsar, and then Stalin, toppled from the ranks of the saints.[15]

In scientific work, too, names may have to be changed by revolutionary acts, once they get established in some arbitrary way. In computing, for example, we are stuck with the terms "fixed-point" and "floating-point" arithmetic, when it is in fixed point that the point floats, and vice versa. That a revolution is required to change such things (consider the metric system in France, or the calendar in Russia) indicates the magnitude of their grip on our minds.

To put the Principle of Indifference into operation, we usually rely on mathematical symbols, which take the sting out of words. The first step in testing the consistency of two observers would be to neutralize the form of their observations. We thus give each observation—each pair, in Herrick's case—an arbitrary name. We might let

$$a = \{\text{lawn, distracted}\}$$
$$b = \{\text{lawn, erring}\}$$
$$c = \{\text{lawn, neglectful}\}$$

and so forth. This translation has the further advantage of getting rid of the *substructure* of the observations, when we are not concerned if the scope or grain differs from observer to observer.

For instance, a second poet might have a language in which Dress and Disorder form a single concept called "Dresorder." In his language, there is no such thing as a {lawn, distracted}, but only a "qualg"; no such thing as a {cuff, neglectful}, but only a "rotz"; no such thing as a {shoestring, careless}, but only a "gliggle." We take the structure out of these observations by the translation:

$$x = \{\text{qualg}\}$$
$$y = \{\text{rotz}\}$$
$$z = \{\text{gliggle}\}$$

and so reduce his viewpoint to the same "one-symbol–one-observation." In this way, each symbol in each set represents precisely one observation for that observer.

Once we have reduced observer A (Herrick) to the range:

$$(a, b, c, \ldots)$$

and observer B to the range:

$$(x, y, z, \ldots)$$

the question of consistency can be easily answered. A is said to be consistent with B if *he never gives two different symbols for one of B's symbols.*

Suppose A and B are watching birds. Every time B says he sees a road runner, A says it is a cuckoo—that much is clearly consistent, since a road runner is a *kind* of cuckoo. Even if B says he sees a yellow-bill, and A *still* says he is observing a cuckoo, that is still consistent, for a yellow-bill is *another* kind of cuckoo. A simply does not, or cannot, make as fine grained a discrimination of birds as B.

If A is consistent with B, we can always tell what A is going to say once we hear B's identification. B says roadrunner, A says cuckoo. B says yellow-bill, A *again* says cuckoo. B says falcon, A says hawk. Figure 3.5 shows this relationship in several ways. First a diagram connects B's observations with A's using arrows, then a tabular form shows the same *mapping* of what B says *onto* what A says. When we try to construct the *inverse mapping,* however, we find that we *cannot predict* what B will say when A says cuckoo.

Mathematically, we characterize this situation by saying that there is a *many-to-one mapping* from B onto A, but a *one-to-many mapping* from A onto B. Since one symbol of A's may map into any one of several of B's, B is *inconsistent* with A, even though A is consistent with B.

Since A is consistent with B, his observations add nothing to those of B. The poet, Herrick, was able to make numerous discriminations in a situation where another man would only observe "Look at her messy clothes!"—which is why the one is a poet and the other a clod. We can dispense with the clod if we have a poet, for the poet *dominates* the clod, as an observer.

Under most general circumstances, neither of two observers will dominate the other in this way. In Figure 3.6 we see a case neither A nor B dominates. We sometimes learn things from A that we could not from B, and vice versa. An interpretation of this situation is shown in Figure 3.7, where A and B are imagined to be looking at a table, one from one side and one from an adjacent side. Because the table is at eye level, if we toss a penny on it, each will be able to tell whether it is to his left or right, but not how close it is. In addition, each can tell whether the penny is on or off the table.

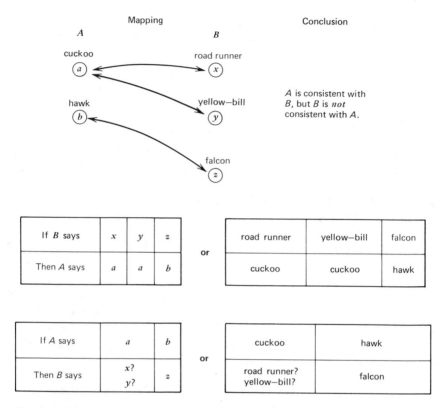

Figure 3.5.   One observer dominating another.

To *A*, there are three observations in this range:

$$a = \text{penny off the table}$$
$$b = \text{penny to my left}$$
$$c = \text{penny to my right}$$

B's range is also three observations:

$$x = \text{penny off the table}$$
$$y = \text{penny to my right}$$
$$z = \text{penny to my left}$$

If we toss a penny towards the table, *A* and *B* will be able to agree if it happens to fall off, though *A* will call that condition *a* and *B* will call it *x*. Anywhere *on* the table, however, we shall not be able to predict what *A* will say from what *B* says, or vice versa. If we can use their information properly, each will make a contribution to our understanding of just where the penny lies.

In this discussion, we have been assuming a special position for ourselves, a point of view that is labeled "our view" in Figure 3.7. It is very easy to slip into imagining that we can somehow get "above the table" when talking about other people's viewpoints, but we really have no reason to believe that we have such super powers of observation. For simple cases, however, we can talk about different points of view if we are willing to introduce an explicit fiction—the "superobserver." It will not be necessary to endow our superobserver with omniscience, but only with a viewing capacity dependent on the abilities of the other observers present. In Figure 3.7, for example, the superobserver would have to be able to discriminate 5 conditions:

$$[(a,x), (b,y), (b,z), (c,y), (c,z)]$$

while for Figure 3.5, he would only have to have the powers of B, since B dominates A.

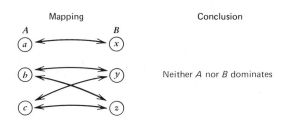

Conclusion

Neither A nor B dominates

If B says	x	y	z
Then A says	a	b or c?	b or c?

If A says	a	b	c
Then B says	x	y or z?	y or z?

Figure 3.6. Two inconsistent observers.

In fact, we can define our superobserver's capacities precisely if we say that *his view must dominate the view of every other observer present.* In the extreme case, this dominance can be *assured* if the superobserver's set of observing states is the *Cartesian product* of all of the others, as shown in Figure 3.8. Why? Because the product set covers *all possible combinations* of the other observations, which is the property that makes us like the Cartesian product.

As before, the Cartesian product is the *largest* case we have to consider. In Figure 3.7, only 5 of the 9 members of the product set were needed, and in Figure 3.5, a superobserver would only need 3. But if we want to be perfectly general and know nothing about the individual observers except their range, we must allow for this maximum case.

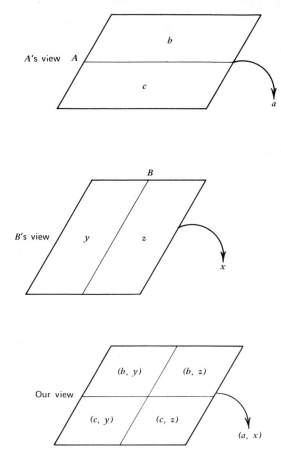

Figure 3.7.   Two points of view—and a third.

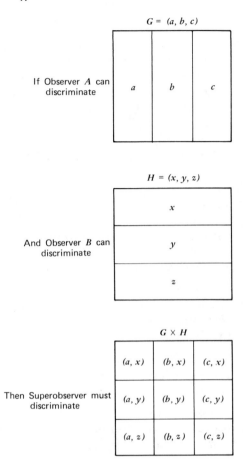

Figure 3.8. The combinatorial powers of a superobserver.

Note that this concept introduces a combinatorial element: the superobserver's powers, though always finite, must grow *much* faster than those of the other observers present. Thus, if 2 observers are present, each capable of discriminating 10 situations, the superobserver must be potentially capable of discriminating 10 *times* 10, or 10^2. Add a third observer and the number grows to 10^3, or 1000. In other words, as we add more and more observers, the *exponent grows* in the discriminatory powers of the superobserver.

Combinatorial growth is a critical flaw in any discussion of multiple points of view, for though we can imagine that a superobserver might exist in simple situations, there is little chance of having one in situations of even modest complexity. We may *discuss* simple situations using the artifice of the superobserver, but we must not imagine that a superobserver can exist in any practical sense. We must particularly refrain from imagining that *we* are the superobserver, capable of seeing what ordinary mortals cannot. Otherwise, *A* will say that *we* are a cuckoo!

QUESTIONS FOR FURTHER RESEARCH

1. *Children's Games*

The requirement that the observer be able to recognize the "same" state when it reappears seems easy enough, but our confidence is shaken when we see the type of game shown in Figure 3.9. Try to solve the puzzle and then, using your experiences, tell what you have learned about observers and observations.

2. *Sociology*

Social scientists are particularly prone to defining sets by implied rules that may be ambiguous to the reader. Consider the following statement by Philip Slater:

Like so many of the more successful nineteenth century utopian communities (Oneida and Amana, for example) the puritans became corrupted by involvement in successful economic enterprise. . . .

Investigate the question of the size of the implied set ("like so many"), the rule for choice ("more successful nineteenth century utopian communities"), and the examples as typical members (Oneida and Amana).

References: Philip Slater, *The Pursuit of Loneliness.* Boston: Beacon Press, 1970.

John Humphrey Noyes, *History of American Socialisms.* New York: Dover, 1966.

3. *Sets*

Give at least five plausible next members for each of the sets given below:

$$\{1, 2, 3, \ldots\}$$
$$\{Mathew, Mark, \ldots\}$$
$$\{pain, gorge, face, \ldots\}$$

4. Pharmacology

Drugs are sometimes relabeled when their "side effects" prove to be more interesting than their main effect. The history of psychoactive drugs is replete

«Il existe 10 différences entre ces 2 dessins: les-
quelles?»

Figure 3.9. There are ten differences between the two pictures: What are they? A popular game in Europe, this example being taken from *Femina,* 25 June 1971 (10, Rue du Valentin, 1004 Lausanne, Switzerland).

81

with such examples. Phenothiazine was initially used as a urinary antiseptic and chlorpromazine was then used to induce artificial hybernation to facilitate anesthesia during surgery. . . . Only later were its psychoactive properties identified as its main attribute. The discovery of the "specific" effects of lithium, amphetamines, iproniazid, and others have similar histories.

> *Reference:* Henry L. Lennard, et al., "Hazards Implicit in Prescribing Psychoactive Drugs." *Science,* **169,** 438 (1970).

Discuss the concept of "side effect" and "main effect" in terms of relativist–absolutist thinking.

5. *Physics—Theory of Elasticity*

Discuss the following statement in terms of relativist–absolutist thinking:

> In general, the deflections in an elastic structure may be said to be "small" if, and only if, they are determined with sufficient accuracy by the classical linear theory of elasticity.

6. *University Life*

In China, according to reports, the universities no longer extrude "academic failures." Discuss what effect this change may have on the role of the university in society—in China, and also (hypothetically) in the United States.

> *Reference:* E. Signer and A. W. Galston, "Education and Science in China." *Science,* **175,** (1972).

7. *Demography*

A village is a system often studied by anthropologists and sociologists. One aspect of a village system is the "set of people living in the village." Discuss how this set might be enumerated, and what practical and conceptual difficulties you might encounter.

8. *Law*

Discuss the analogy between an arbitrator or judge and a superobserver.

9. *History as Observation*

Let us suppose that a military commander has just won a victory. That, immediately, he sets to work writing an account in his own hand. That it was he who conceived the plan of the battle, and that it was he who directed it. And finally that, thanks to the moderate size of the field (for in order to sharpen the argument, we are imagining a battle of former times, drawn up in a confined space), he has been able to see almost the entire conflict develop before his eyes. Nevertheless, we cannot doubt that, in more than one essential episode, he will be forced to refer to the reports of his lieutenants. In acting thus as narrator, he would only be behaving as he had a few hours before in the action. Then as commander, regulating the movements of his troops to the swaying tide of battle, what sort of information shall we think to have served him best? Was it the rather confused scenes viewed through his binoculars, or the reports brought in hot haste by the couriers and aides-de-camp? Seldom can a leader of troops be his own observer. Meanwhile, even in so favorable a hypothesis as this, what has become of that marvel of "direct" observation which is claimed as the prerogative of the studies of the present?

In truth, it is scarcely ever anything but a delusion, at least as soon as the observer has expanded his horizon only slightly. A good half of all we see is seen through the eyes of others.

Discuss how much of the "observation" in your own discipline is "through the eyes of others."

Reference: Marc Bloch, *The Historian's Craft,* p. 49. New York: Vintage Books, 1953.

10. *Schools and the Banana Principle*
One of my students, Jim Addiss, made this comment about an application of the Banana Principle:

I decided to go back to school, but I don't yet know how to stop.

Another (anonymous) student commented:

Students are taught in school the method of doubting, but they never are taught when to stop, so they wind up committing suicide.

Comment on these comments, in the light of the Banana Principle; give a few examples from your own school experience of how schools obey the principle; and make some suggestions as to how schools could teach when (or how) to stop applying what they teach.

READINGS

RECOMMENDED

1. Authur, D. Hall and R. E. Fagen, "Definition of System." In *Modern Systems Research for the Behavioral Scientist,* Walter Buckley, Ed. Chicago: Aldine, 1968.
2. Eleanor Gibson, "The Development of Perception as an Adaptive Process." *American Scientist,* **58,** 98 (January–February 1970).

SUGGESTED

1. E. H. Gombrich, *Art and Illusion.* New York: Pantheon Books, 1961.
2. Studs Terkel, *Hard Times.* New York: Avon Books, 1971.

NOTATIONAL EXERCISES*

1. Write down the set of all "first words of a line" from Herrick's poem. How many elements are in this set?

2. Take the set from Exercise 1 and divide it into *subsets* according to the first letter of the word. That is, all words in each subset should start with the same letter.

3. Take the set from Exercise 2 and give each subset a symbol—the symbol being the first letter of each word in the subset. Write down the set of all these symbols. Describe in words what it is a set of.

4. Repeat Exercises 2 and 3, but based on the *last* letter of the *last* word in each line.

5. Assume we have two observers, "First-letter" and "Last-letter," who can only see, respectively, the first and last letter of each line of Herrick's poem. Write down the *product set* that a superobserver would have to be able to resolve to guarantee being a superobserver relative to this pair of observers.

6. Does the superobserver in Exercise 5 actually have to resolve all pairs in this product set? Explain your answer.

7. Suppose we have a third observer, "Odd–even," who can only discriminate, by some means or another, whether a line is an odd or even line of a poem, so that his grain is the set $\{O,E\}$ corresponding to lines $\{1, 3, 5, 7, 9, 11, 13\}$ for O and $\{2, 4, 6, 8, 10, 12, 14\}$ for E. Is Odd–even *dominated* by either First-letter or Last-letter? Write down the mappings needed to determine the answer.

8. Does the superobserver of Exercise 6 have to expand his powers in order to *dominate* Odd–even?

ANSWERS TO NOTATIONAL EXERCISES

1. $\{A, Kindles, Into, An, Enthralls, Ribbands, In, I, Do, Is\}$ Notice that in set theory, known duplicates are conventionally discarded from the set, so that A only appears once, so there are only 10 elements, even though it is a sonnet form of 14 lines.

2. $\{(A, An), Kindles, (Into, In, I, Is), Enthralls, Ribbands, Do\}$
Technically, since each element is a set, single words should also be

---

* Whenever new notation is introduced, exercises will be supplied for the convenience of the reader who wishes to practice. Since mathematically trained readers will probably not need to do these exercises, they are kept separate from the Research Questions. They are, moreover, not research questions, but simply practice in notation. The reader should check his work against the answers, to verify his mastery of notation.

written with parentheses, such as (Kindles), but we will avoid this detail unless it is needed for clarity.

3. {A, K, I, E, R, D}
This is the set of all initial letters of lines in Herrick's poem.

4. {dress, wantonness, thrown, distraction, there, stomacher, thereby, confusedly, note, petticoat, tie, civility, art, part}
{(dress, wantonness), (thrown, distraction), (there, note, tie), stomacher, (thereby, confusedly, civility), (petticoat, art, part)}
{S, N, E, R, Y, T}
This is the set of all final letters of lines in Herrick's poem.

5. { (A,S),   (A,N),   (A,E),   (A,R),   (A,Y),   (A,T)
     (K,S),   (K,N),   (K,E),   (K,R),   (K,Y),   (K,T)
     (I,S),   (I,N),   (I,E),   (I,R),   (I,Y),   (I,T)
     (E,S),   (E,N),   (E,E),   (E,R),   (E,Y),   (E,T)
     (R,S),   (R,N),   (R,E),   (R,R),   (R,Y),   (R,T)
     (D,S),   (D,N),   (D,E),   (D,R),   (D,Y),   (D,T)}

6. Given this particular poem, the superobserver actually needs only to be able to discriminate the 11 pairs:

{(A,S),   (K,S),   (A,N),   (I,N),   (A,E),   (E,R),   (A,Y),
(R,Y),   (I,T),   (I,Y),   (D,T)}

because first of all, there are only 14 lines to the poem, so he couldn't possibly have to discriminate more than 14 states. Second, 3 lines correspond to {A,E} and 2 to {I,T}, which eliminates the need for three discriminations, since the pair of observers together can only discriminate 11 lines.

7. This observer is dominated by First-letter but not by Last-letter. The mapping from First-letter onto Odd–even is the following:

First says	A	K	I	E	R	D
Odd–even says	*O*	*E*	*E*	*E*	*E*	*O*

which is perfectly predictable, because of the highly odd–even pattern of first words.

    The best mapping we can do from Last-letter to Odd–even is:

Last says	S	N	E	R	Y	T
Odd–even says	?	?	*O*	*E*	?	?

because rhyme pairs on adjacent lines tend to have the same last letter.

8. No, because if superobserver already dominates First-letter and First-letter dominates Odd–even, then superobserver will also dominate Odd–even. As we shall see, "dominance" is what we call a "transitive" property.

4

Interpreting Observations

The village idiot had just won the annual lottery, with a prize of two handsome horses and a fine carriage. One of the ne'er-do-wells of the village hitched a ride with him on his new carriage and asked, "How did you manage to pick the winning number, anyway? What's the secret of your success?"

The slow-witted one, not seeing the barb in the query, replied, "Oh, it was easy. You see, my lucky number is seven, and the lottery was held on the seventh of the month, so I multiplied seven times seven and got sixty-three, the winning number."

"You idiot!" the loafer laughed, almost falling out of his seat. "Don't you know that seven times seven is forty-nine?"

"Oh," said the idiot, seeing at last that he was being mocked, "you're just jealous."

Traditional Story

States

A state is a situation which can be recognized if it occurs again.

Author Unknown

Having concluded the last chapter with a stern warning, let us now ignore that warning for the duration of the following discussion of observation. Imagine that you walk into a strange room containing a *big black box*. Since there are no other observers currently in the room, we shall have to assume that you are not only a superobserver, but a super-superobserver. That is, we shall have to pretend *that no matter what observers* ultimately come into the room, you will have sufficient observational powers to dominate them.

Actually, there will be only two other observers, and they will be along presently. Notice, however, that the concept of super-superobserver is much like the concept of "reality," in that it contains "all possible" observations. Put in these terms, we see that the concept of "reality" is very close to what some people call "God."

87

Well, here you are, playing God in a room with a black box. Since you have super-superpowers, you note immediately that the only things worth observing on the box are a red light (R), a green light (G), and a whistle (W), which we take to be the *scope* of your observation and label

$$S = \{R, G, W\}$$

The lights may be either on or off: we say that they each have two possible "states." Just to practice your new-found virtuosity with notation, you decide to identify an "on" state with the number 1 and an "off" state with the number 2. These numbers have no value as *measures,* but are simply names, like x, a, S, or Katz.

The *range* of your two light observations is therefore

$$R = (1, 2)$$
$$G = (1, 2)$$

The whistle is a bit more complex, for it has six tones, which you characterize as

$$W = (1, 2, 3, 4, 5, 6)$$

Since the scope of your observations is

$$S = \{R, G, W\}$$

we can immediately write down all possible states by taking the Cartesian product, as shown in Figure 4.1. Since you are about to observe the *behavior* of the black box, you abbreviate the elements of the product set:

$$a = (1, 1, 1)$$
$$b = (1, 1, 2)$$

and so forth, as shown in the Figure.

This shorthand will help you record the behavior of the box, for though you have super observing powers, you don't have a super memory. You take out a pencil and paper and write down the *sequence of observations,* which happens to be

$$\dots a \; n \; i \; k \; a \; n \; i \; k \; a \; n \; i \; k \; a \; n \; i \; k \; a \dots$$

This sequence of symbols is a lot shorter than writing

The two lights are on and there is a low whistle, then the red light goes out and the whistle rises one step, then the tone rises one more step while the red light goes on and the green light goes off, then the lights stay the same but the tone jumps two steps, then the green light goes on and the tone falls back to its lowest point . . .

though you may verify that it says the same thing.

R	G	W	Name
1	1	1	a
1	1	2	b
1	1	3	c
1	1	4	d
1	1	5	e
1	1	6	f
1	2	1	g
1	2	2	h
1	2	3	i
1	2	4	j
1	2	5	k
1	2	6	l
2	1	1	m
2	1	2	n
2	1	3	o
2	1	4	p
2	1	5	q
2	1	6	r
2	2	1	s
2	2	2	t
2	2	3	u
2	2	4	v
2	2	5	w
2	2	6	x

Figure 4.1. All possible states of the black box.

You are lucky that there is a great deal of regularity, or *constraint,* in the sequence, otherwise you might have a lot more writing to do. As usual, it may be wise to ask, "How much more writing?" to appreciate the magnitude of the problem. We cannot simply ask "How many possible sequences are there?" Since a sequence could be indefinitely long, there are an infinite number of possible sequences. What we *can* ask is "How fast does the number grow as the length of the sequence grows?"

If there are two observations in a sequence, the sequence is a pair of choices from the set of 24 states. All possible pairs is thus the *product set,* which has 24^2 members. Therefore, there are 24^2 (576) possible sequences of length 2. Similarly, there are 24^3 (about 14,000) sequences

Ordered pairs

(a, n)
(n, i)
(i, k)
(k, a)

Mapping

State		a	n	i	k	from (a, n, i, k)
Is followed by State		n	i	k	a	into (a, n, i, k)

Directed graph

Figure 4.2.   Three representations of the same sequence.

of length 3; 24^4 (about 300,000) of length 4; or, in general, 24^n sequences of length n. In other words, the number of possible sequences grows *combinatorially* with the length of the sequence.

A superobserver will require a particularly super memory if he is to remember everything he sees—or else he will have to be very lucky and see only highly *constrained* sequences. Because our sequence is highly constrained, we can utilize some very compact means of recording what we saw, as illustrated in Figure 4.2. In the first place, we can write down each and every predecessor–successor pair, which gives us some idea of the *degree* of constraint. Out of the 576 possible *ordered* pairs, only 4 actually occur.

Because there are so few pairs, we can easily translate them into the form of a table that is a *mapping* from the set of observed states *into itself* (Figure 4.2). We used the *form* of this mapping in the previous chapter to designate a correspondence between one point of view and another. A mapping, however, can express *relationship* between *any* two sets—or, more precisely, *from* any set *to* any other—including itself.

In this case, then, the mapping indicates sequence, which is another form of relationship: the relationship between one observation and the observation that follows it. Since there is no ambiguity in this mapping—that is, since it is *not* one to many—the behavior of this sequence is *perfectly predictable* after just one observation. This predictability becomes somewhat more evident if we use the *directed graph* notation showing arrows going from each state to its successor (Figure 4.2). Each arrow thus corresponds to one of the *ordered pairs* or one of the *entries* in the mapping table.

Although each of the three forms of representation in Figure 4.2 is mathematically equivalent to the others, they are not equivalent from a psychological point of view. In the directed graph form, for example, we can see quite readily that the sequence forms a *cycle*—something not so evident in either of the two others.

Another way you might have recognized a cycle is by being *bored,* if superobservers had any feelings. You do not know, of course, that the sequence might not change at any instant, but after a few hundred repetitions, you make the inductive leap: "My God, it will continue like this *forever—unless I do something."*

Up until now, you have been a completely *passive* observer. Although you are omniscient (all seeing) you are not omnipotent (all powerful). In fact, as superobserver, you have no power at all—you are omniscient and impotent at the same time.

You have been playing a game that systems researchers call "black box." The rules of the black-box game require that the observer be impotent to manipulate the box by looking "inside." The object of playing this conceptual game is to furnish insight into the process of observation. The black box can be applied as a conceptual tool[1], or as an effective teaching tool[2], but we must not mistake it for a serious model of many real observers.

What the black box does model is an observer who cannot or will not influence the system that he is investigating. It models rather well an astronomer studying the expanding universe, but breaks down as we get closer to home. When, for example, a hairy black box about the size of a teacup crept on its eight furry legs out of the yarn basket, I was content to treat it as a black box. But Heathcliff, a fearless German shepherd, had his own way of investigating our arachnid friend. He watched a bit, sniffed a bit, and then cuffed it first in one direction and then in the other with his paw. Using the paw is against the rules of black box, but Heathcliff does not know that game. He had to find out if the black box wanted to play or be eaten.

When they have the courage, human beings also interact with the systems they observe. Even Daniel F., sitting with his can of beer in front of the TV, retains the power to change channels when he gets bored. We may believe the world to be independent of the *percipient* observer, but we definitely feel it depends on the *participant* observer.

If you can change things, you may not be so bored, so before the black box loses your attention, let us suspend the rules of the simple black-box game and endow you with certain very limited powers of interaction.

First you begin to examine in detail its Edwardian decoration. Quite by accident, you touch a concealed spring and a little door swings open on one side. Inside the door is a sign:

KICK ME

You remember what happened to Alice when she found the bottle labeled

DRINK ME

but are more afraid of ennui than of fear. Like Heathcliff with the spider, you give the box a tentative tap on the baseboard.

Instantly, the pattern of light and sound changes, and we see

. . . *g m d f g m d f g m d f g* . . .

This cycle holds your interest for a short time, after which you give it a bolder kick, with this result:

. . . *b j r c q h p l o e b j r* . . .

Being a bit longer, this cycle holds your interest longer, but when you finally resume kicking, you fail to get any other behaviors than the three cycles already found, which you plot out as in Figure 4.3. Although there are 6 possible states that you have never seen, you eventually give up hope of ever seeing them. Discouraged, you leave the room in search of some marinated artichokes, for even superobservers get hungry after a long session of observation.

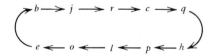

Figure 4.3. Directed graphs of the three cycles of the black box.

But your appetite is not yet to be satisfied, for as you leave the room you encounter a friend who asks, "What's going on in there?"

"Not very much," you reply, stifling a yawn and trying to conceal the growling in your stomach. "Only some kooky nineteenth-century contraption. It's got interesting decorations, but its behavior is rather a bore—just three cycles, and all perfectly determinate. You have to kick it to get it to change cycles, but it seems harmless."

"Sounds odd," he says. "If you wait here while I take a peek, then we can go have lunch together."

The promise of lunch keeps you waiting a rather long time, and when your friend emerges, he has a look of half bewilderment and half pity on his face. "It's actually rather cute, but I'm afraid you're not a very good observer."

"What do you mean, 'not a very good observer.' I'm a super-superobserver."

"Well, first of all it has only two cycles, not three. And second, it's not determinate at all."

"That's simply not true," you reply. "It has two four-state cycles and one ten-state cycle. I watched it for an hour and a half."

Just then a stranger in a tall beaver hat strides up and interrupts. "You're right about the three cycles, but the longest cycle is only five notes long."

"How do you know? You weren't even in the room."

"I built it. Who should know better than I about my own music box?"

"Music box? That's no music box."

"If I say it's a music box, that's what it is. A royal music box, at that—a gift for the Red Queen. . . ."

"Look here," you interrupt. "Royalty be hanged. The author of this book told me I'm a superobserver—a super-superobserver, in fact—so if I say it's not a music box, then . . ."

"Don't interrupt! Just because you're a see-it-all doesn't mean you're a know-it-all."

"Hey, you two," your friend shouts, stepping between you just as fists begin to rise. "Let's be *rational*. *You* may be the inventor, and *you* may be a superobserver, but I have a Ph.D. in Physics. I know how to make observations, so let's just go inside and I'll prove to you that you're *both* wrong. It obviously has only two cycles."

"If it does, then you must have broken it with all that kicking. Who said you could kick my machine, anyway?"

"Come off it! The machine itself said 'Kick me.'"

All three of you file through the door, whereupon the inventor gives a little speech. "You see, 'Kick me' is the Queen's favorite expression, and means 'Yell at me.' The music box plays the three national anthems, and to get it to change its tune, all she has to do is yell, which she adores doing. Watch!" he shouts, and the pattern changes. After a few moments, another shout makes it change again, at which point everyone declares in unison:

"See, it works just the way I said."

The Eye–Brain Law

I cannot erase from my mind a certain film supplied graciously and gratuitously to me by the French consulate for use in a class many years ago. I was teaching a survey of the cultures of Southeast Asia, and the film showed the temples of Angkor. In one segment, running perhaps a couple of minutes, an old, white-bearded, and very dignified, if anonymous, professor was shown conducting another seemingly academic person about a portion of the ruins. A fallen piece of sculpture attracted the older man's eye and he brought it to the attention of his companion. Unfortunately, at that very moment a Khmer workman was bending in front of the object, possibly at some task. Without thought or hesitation, the old archaeologist poked the Cambodian away with a long fan he was carrying. So commonplace was the event that it obviously never occurred to anybody in the French information service to excise that bit of colonialist by-play.

Morton H. Fried[3]

After your first experience of playing superobserver, you should never again be surprised to see something that others—even in the French information service—do not see. By applying the Principle of Indifference to "you" and "others" in the previous sentence, we get another insight that is always a little harder to accept. Perhaps if we trace the dialogue a bit further, we will understand why you and the inventor almost got into a fight.

The inventor explains to us that the music box plays six different notes, which we already knew but which is a revelation to our physicist friend, who is slightly deaf and can only hear three tones. When we ask about the lights, the inventor replies, "What lights?"

"The lights on the front—the red one and the green one?"

"Oh, they're *nothing*. As long as either one of them is on everything is okay. It's just a safety device—has nothing to do with the music box."

"You know," our physicist friend chimes in, "you shouldn't use a red light on a piece of equipment unless it means danger. I always follow that safety practice in my lab. I knew the green light was unimportant, but red lights have a special meaning on equipment. You really should fix that, you know."

Suddenly the mystery is clear. Since the inventor ignores the lights, he sees only six states—the six tones of the whistle. To him, the "purpose" of the box is known, which means that he does not have to discriminate as many states as you, the superobserver, did. As Figure 4.4 shows, each of his states corresponds to *four* of ours, which has the effect on behavior seen in Figure 4.5.

Because of the way the inventor "lumps" our states into his, we could apply a mapping to the directed graphs of your superview and obtain the inventor's view. Thus, for example, the cycle *"anik"* is transformed into the cycle *"ABCE,"* If your view did not *dominate* the inventor's view, we should be unable to make this mapping in a unique way, just as he is unable to map his view into yours.

Although each of your states maps into one state of his, the structure you see is different. For one thing, what you saw as a cycle of 10 states becomes for him a cycle of five states, *"BDFCE,"* traversed twice—just as a school year to the janitor is two semesters to the registrar.

Another difference is that your view is "state determined" whereas his is not. He cannot draw a mapping as we did in Figure 4.2 because each state is not *always* followed by the same state. In his second cycle, for example, he would see both the pair (A, A) and the pair (A, D). Moreover, his state A also appears in the first cycle followed by B, so that if he comes to the machine fresh and observes one state, he cannot be *sure* which state will come next.

Notice, however, that *if the inventor can remember the previous two states*, he would be able to predict the next state. A may be followed by B, D, or A itself, but the *sequence* (F, A) can only be followed by A. Such substitution of mental capacity for observing power is an illustration of a general law about observers, which we may call *The Eye–Brain Law*:

State	Tone	Our corresponding states
A	1	(a, g, m, s)
B	2	(b, h, n, t)
C	3	(c, i, o, u)
D	4	(d, j, p, v)
E	5	(e, k, q, w)
F	6	(f, l, r, x)

Figure 4.4. The inventor's point of view.

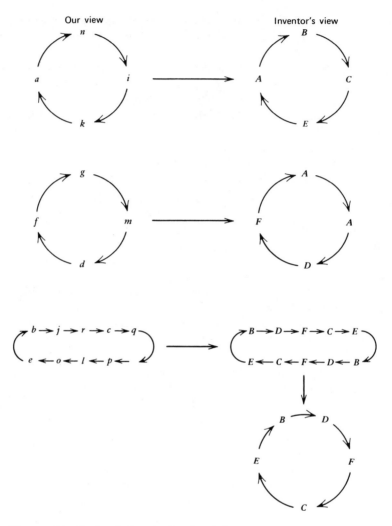

Figure 4.5.   Seeing the inventor's point of view.

To a certain extent, mental power can compensate for observational weakness.

Through symmetry, we can immediately derive *The Brain–Eye Law*:

To a certain extent, observational power can compensate for mental weakness.

Note that the friend (whose different view is given in Figures 4.6 and 4.7) would need *greater memory* to overcome his observational inade-

State	Red	Tone	Our corresponding states
S	1	(1,2)	(a, b, g, h)
T	2	(1,2)	(m, n, s, t)
U	1	(3,4)	(c, d, i, j)
V	2	(3,4)	(o, p, u, v)
W	1	(5,6)	(e, f, k, l)
X	2	(5,6)	(q, r, w, x)

Figure 4.6. The friend's point of view.

quacy, since if he sees (V, W) it might be followed by either S or V. Figure 4.6 shows that his observational power has a range of 6 states, as does the inventor's. The *amount* of mental power must evidently depend on something *other than the mere number of states* discriminated.

Many examples of the Eye–Brain Law spring to mind. The experienced doctor needs fewer laboratory tests than the intern to

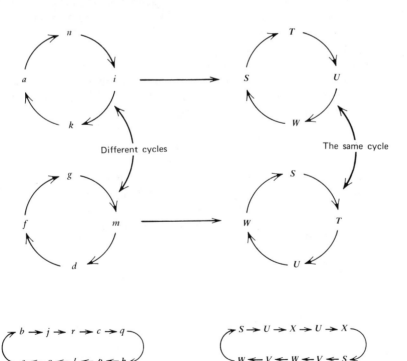

Figure 4.7. Seeing the friend's point of view.

make the same diagnosis, but to some extent the intern can substitute a good laboratory for the years of experience he has not had time to accrue. We drive more slowly at night to give us more time to observe potentially dangerous situations, in compensation for our reduced vision. We write ourselves notes on scraps of paper for later reading, in order to relieve our overburdened memory.

The Eye–Brain Law will *not* work in any case where there is no constraint whatsoever in the observation. Clearly, memory is of no use unless the future is like the past. The physicist was led astray about the green light and the red light because in his past, machines did not use lights in that way. He lumps too much to see a state-determined system, but he discriminates too much to recognize that the black box is "really" a music box.

The superobserver, since he is prepared to make the maximum possible discriminations, is effectively operating without memory. Although he will be sure to see the fine details, he may easily miss the big picture. As the inventor remarked, "see-it-all" doesn't mean "know-it-all," for knowing means knowing how to ignore certain details. We can only "learn" by seeing the "same" situation repeat itself. This is what we mean by "state"—a situation that our observer can recognize if it occurs again.

Discriminating too many states is what we have previously called *undergeneralization.* The popular image of science envisions the scientist making the maximally precise measurements as a basis for his theories, but, in practice, scientists are lucky that measurements are not overly precise. Newton based his Law of Universal Gravitation on the elliptical orbits of Kepler, but Kepler abstracted these ellipses from the observations of Tycho Brahe. Had those observations been a bit more precise (as precise as we now can make) the orbits would not have been seen as ellipses, and Newton's work would have been much more difficult. With *more precise* observations, the simplifications we discussed in Chapter 1 would have been left for Newton to make *explicitly*—thus immensely compounding his difficulties.

Thus, the balance between "eye power" and "brain power" cannot be pushed too far in either direction. The problem of science is to find the appropriate compromise.

The Generalized Thermodynamic Law

When asked, for example, what happens to two blocks of copper initially at different temperatures left alone together in an insulated container, they will all reply that the blocks will come to the same temperature. Of course, if asked how they know, they usually say "Because it is a law of nature.". . . the opposite is true . . . it is a law of nature because it happens."

John R. Dixon and Alden H. Emery, Jr.[4]

The divergence of views about the music box indicates that the observational problem may get greater when systems of greater complexity are involved. We can imagine situations, however, in which these three observers would all see the same thing. For instance, if the actual behavior of the box was such that the lights simply stayed fixed and the only tunes played involved notes (1, 3, 5), the three views would coalesce. The physicist cannot hear the difference between 1 and 2, but since 2 never sounds, his tone deafness is not a handicap. Moreover, since the lights never change, you, the superobserver, will not observe as many possible states, just as you never observed states (s, t, u, v, w, x) because at least one light was always on.

Figure 4.8 illustrates some behaviors that this system might exhibit in which these three observers would agree, in spite of their rather different observational powers. While what we observe depends on our characteristics as observers, it does not depend *entirely* on those characteristics. There are two extreme views on this subject—the "realist" and the "solipsist." The "solipsist" imagines that there is no reality outside of his own head, while the "realist" flatters himself that what is in his head is all real. Both suffer from the same disease.

There are two components to any observation: a duality long known and often forgotten. Galileo

. . . distinguished between primary qualities of matter and secondary qualities—the former inherent in matter itself, the latter the product of the interaction of the body possessing certain primary qualities with the sense organs of a human or animal observer.[5]

Many of Galileo's intellectual heirs have forgotten this distinction, making it appear prudent for us to frame a law, which we shall call *The Generalized Thermodynamic Law*:

More probable states are more likely to be observed than less probable states, unless specific constraints exist to keep them from occurring.

In spite of the risk of antagonizing physicists, we call this law the Generalized Thermodynamic Law because it has two important parts that correspond to very general cases of the First and Second Laws of Thermodynamics. The First law, we recall, concerns the conservation

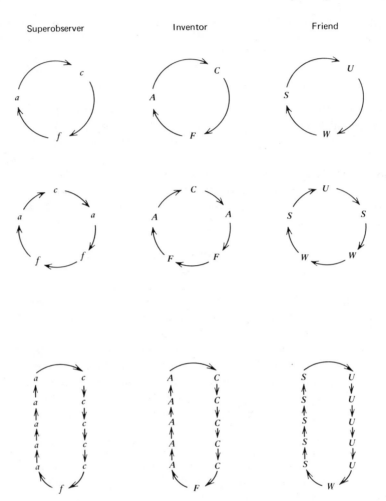

Figure 4.8.   The music box playing three different tunes.

of something called "energy," which seems to conform to a rather severe constraint "out there." The Second Law, however, is of a different type, being concerned with the limited powers of observers when viewing systems of large numbers of particles. By analogy with these laws and with Galileo's primary and secondary qualities, we may reframe our law:

> *The things we see more frequently are more frequent:*
>   *1. because there is some* **physical** *reason to favor certain* **states** (the First Law)

or

2. *because there is some* **mental** *reason* **(the Second Law).**

Because of the predominance of realist beliefs, it is hardly necessary to elaborate on 1, though there is more to it than meets the eye. Our purpose in propounding this law is to correct the tendency to overdo realist thinking—to the point of stifling thought. We shall therefore illustrate the Second Law with an extended example.

Which of the two bridge hands in Figure 4.9 is more likely to be seen in a normal bridge game? (You really need know nothing about bridge—we are speaking essentially of dealing 13 cards in an honest deal.)

Most bridge players readily answer that Hand 2 is the more likely, but statisticians tells us that the two likelihoods are the same. Why? In

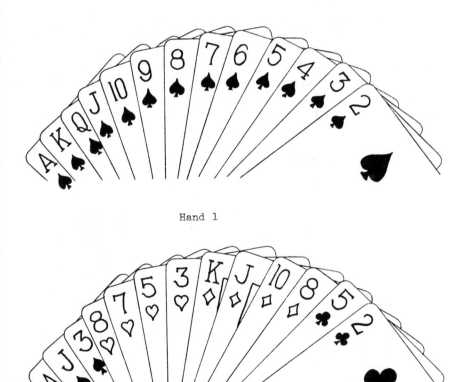

Hand 1

Hand 2

Figure 4.9. Two bridge hands.

an honest deal, *any precisely specified* hand of 13 cards has to be as likely as any other hand so specified. Indeed, that is what statisticians *mean* by an "honest deal," and it also agrees with our general systems intuition based on the Principle of Indifference. What should the deck care what names are painted on its faces?

But the intuition of bridge players is different. We want to know why they instinctively pick Hand 2 to be more likely than Hand 1. The reason must lie in the structure of the game of bridge, the arbitrary rules that attach significance to certain otherwise indifferent card combinations.

When we learn a card game, *we learn to ignore certain features* that for that game are not usually important. Games such as War and Old Maid where suits are not important may be played without taking any notice of suits whatsoever.

In bridge, *high* cards are usually important, and cards below 10 are usually not important. We are always impressed when a bridge expert makes a play involving the "insignificant" five of hearts. Bridge books implicitly recognize the unimportance of low cards by printing them as anonymous x's, just as the government implicitly recognizes our unimportance by reducing us to unnamed statistics. In a typical lesson, we might see Hand 3 of Figure 4.10. Seeing this hand, we may understand the bridge player's difficulty by asking the analogous question:

Which is more likely, Hand 1 or Hand 3?

Though bridge players are only dimly aware of the fact, Hand 3 is not a "hand" at all, but symbolizes a *set of hands*. Hand 2 happens to be one *member* of the set we call Hand 3, and when a bridge player looks at a hand such as Hand 2, he unconsciously performs various lumping operations, such as counting "points," counting "distribution," or ignoring the "small" cards. Therefore, when we show him Figure 4.9 he thinks we are asking

Which is more probable, a hand *like* Hand 1 or a hand *like* Hand 2?

There is, in the bridge player's mind, only *one* hand *like* Hand 1, since it is the only guaranteed unbeatable grand slam in spades, the highest suit. Hand 2, however, is quite "ordinary," and any hand in the set "Hand 3" will be more or less *like* it. Since the set we call Hand 3 contains more than a million individual hands, the odds are at least a million to one in favor of getting a hand *like* Hand 2! Because of his past experience, the bridge player has *translated* our question to another—one to which his answer is perfectly correct.

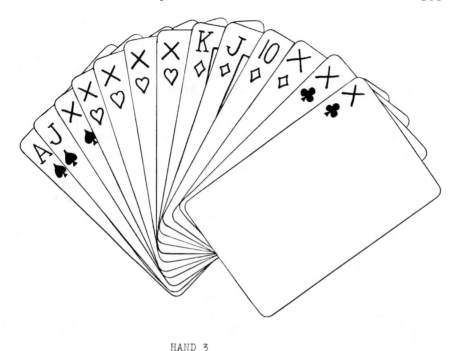

HAND 3

Figure 4.10. A "hand" that is not a hand.

We habitually translate questions in such ways. Consider what the statistician did with our original question, which asked

Which is more likely *to be seen* . . .

Because we ordinarily speak rather casually, he translates "be seen" into "occur," which leads him to an important mistake. In actual bridge games, Hand 1 will be *seen* much more often than ever Hand 3! Why? Because although Hand 3 may *occur* more often, it will rarely, if ever, be *seen*—that is, specifically *noticed* by any of the players. Although Hand 1 may not *occur* very often, if it ever does it is sure to make the morning papers, because it is such a sensational hand in the game of bridge.

The *importance* of hands and the *way they are lumped* are obvious mental reasons for seeing some "hands" more frequently than others. These arguments are based on the assumption of an "honest deal," which is another way of saying that there are *no physical reasons* to favor any particular hand. Those who have played much bridge, though, know that there *are* such physical reasons. If a player leaves the table for a few minutes during a friendly game, he may come back to find he has been dealt 13 spades. If he is not too gullible, he will look up and laugh, because he will know that his friends have rigged the hand as a practical joke.

But how does he *know* they have rigged the hand? Well, because he knows that *in an honest deal,* such a hand is vastly improbable. Yet so is any other hand, but we do not think that every hand comes from a dishonest deal. Whenever we observe a state that is both *conspicuous* and *improbable,* we are faced with a quandary. Do we believe our observation or do we invoke some special hypothesis? Do we call the newspaper, or do we accuse our friends of rigging? Only a fool would call the newspaper; and if he did, no sensible person would believe him.

In the same manner, conservatism is introduced into scientific investigation by the very assumption that observations must be consonant with present theories. An observation is more likely to be simply discarded as "erroneous" if it is out of consonance with theory. If the observation is nonrepeatable, it is therefore lost forever, leading to a selection for "positive" results. Such selection can be seen most vividly when researchers dig up historical data, and especially in connection with old astronomical observations needed to give a time perspective not obtainable in the present. Robert Newton[6] calls these procedures for assigning dates and localities to questionable observations the "identification game," and nicely shows the way the procedures can lead to the "classical" results even if a table of random numbers is used as the "raw" data to be identified.

The complete substitution of theory for observation is, of course, not scientific. Even worse is going through the motions of observing, but discarding as "spurious" every observation that does not fit theory—like the Viennese ladies who weigh themselves before entering Demel's Tea Room. If they're down a kilo, they have an extra mochatorte, and if they're up a kilo they pronounce the scale "in error" and have an extra mochatorte anyway.

This, then, is the problem. Raw, detailed observation of the world is just too rich a diet for science. No two situations are *exactly* alike unless we make them so. Every license plate we see is a miracle. Every human being born is a much greater miracle, being a genetic combination which has less than 1 chance in 10^{100} of existing among all possible genetic combinations. Yet the same is true for any *particular* state—in the superobserver sense—of any complex system.

"A state is a situation which can be recognized if it occurs again." But *no* state will ever occur again if we don't lump many states into one "state." Thus, in order to learn at all, we must forego *some* potential discrimination of states, some possibility of learning everything. Or, codified as *The Lump Law:*

If we want to learn anything, we mustn't try to learn everything.

Examples? Wherever we turn they are at hand. We have a category of things called "books" and another called "stepladders." If we could not tell one from the other, we would waste a lot of time in libraries. But suppose we want a book off the top shelf and no stepladder is at hand. If we can relax our lumping a bit, we may think to stack up some books and stand on them. When psychologists try this problem on people, some take hours to figure out how to get the book, and some never do.

It's the same in any field of study. If psychologists saw every white rat as a miracle, there would be no psychology. If historians saw every war as a miracle, there would be no history. And if theologians saw every miracle as a miracle, there would be no religion, because every miracle belongs to the set of all miracles, and thus is not *entirely* unique.

Science does not, and cannot, deal with miracles. Science deals only with repetitive events. Each science has to have characteristic ways of lumping the states of the systems it observes, in order to generate repetition. How does it lump? Not in arbitrary ways, but in ways determined by its past experience—ways that "work" for that science. Gradually, as the science matures, the "brain" is traded for the "eye," until it becomes almost impossible to break a scientific paradigm (a traditional way of lumping) with mere empirical observations.

Functional Notation and Reductionist Thought

By means of mathematics we purchase a great ease of manipulation at the cost of a certain loss of complexity of content. If we ever forget this cost, and it is easy for it to fall to the back of our minds, then the very ease with which we manipulate symbols may be our undoing. All I am saying is that mathematics in any of its applied fields is a wonderful servant but a very bad master; it is so good a servant that there is a tendency for it to become an unjust steward and usurp the master's place.

Kenneth Boulding[7]

The black-box model of observation exposes certain aspects of the investigative process, but tends to give a rather passive view of the entire matter. The observer is allowed neither to change the box, nor himself. In any real situation, the observer must face the problem of defining his scope and grain of observation. Since these characteristics may have a decisive effect on what he learns, we must not simply pass over this process with a wave of the hand.

When the observer chooses a particular scope of observation, he is, in effect, declaring his belief that those are the important features—or at least the most important of the ones he can observe. There is a fine mathematical shorthand for just this situation, called *functional notation*. If we write, for example,

$$z = f(a, b, x)$$

(which we read, "*z is a function of a, b*, and *x*," or "*depends on a, b*, and *x*"), we are stating explicitly that *z* depends on *a*, *b*, and *x*— and *only* on *a*, *b*, and *x*, as far as we know or care at present. The small letter *f* is conventionally used to mean "some function of," but other letters may be used. If, for instance, we wrote

$$y = g(a, b, x)$$

we would be saying that *y* is *also* a function of *a*, *b*, and *x*; but by using the letter *g* we emphasize that *y* is a *different* function of *a*, *b*, and *x*. That is, though *y* depends on the *same things* as *z*, it depends on them in a different way.

Functional notation is especially important in general systems thinking, for it permits us to express *partial knowledge* about a system when we do not know how to characterize its behavior exactly. Newton, for example, might have said that

$$F = f(M, m, r)$$

before he was able to give exact form to the force of gravitational attraction. The notation states that the force (F) between the two bodies depends (only) on the two masses (m and M) and the distance (r) between them. Once this stage has been reached, the job of finding the exact nature of the functional relationship—the Law of Universal Gravitation—is much easier.

Functional notation can also be mixed with explicit formulas to show an *intermediate stage of knowledge* between that of functional dependencies and exact formulas. Had Newton discovered the inverse square property before knowing about the product of the masses, he might have summarized his knowledge by writing

$$F = \frac{g(m,\ M)}{r^2}$$

that is, the force depends inversely as the square of the distance, and on some *unknown function* of the masses.

Similarly, if our knowledge of the functional dependencies may not be complete, we can use the ellipsis (...) notation within the functional notation, as if Newton had said

$$F = \frac{h(m,\ M,\ ...)}{r^2}$$

meaning that he knew the force varied inversely as the distance, that it depended on the masses in some way, and that something else might be involved that was not as yet known.

Many general arguments can be carried out on the rather low level of specific knowledge implied by a simple functional relationship. Consider our discussion of the Principle of Indifference in which a researcher presented us with some formula.

$$D = f(S, R)$$

where

$$D = \text{difficulty of making a selection}$$
$$S = \text{percentage of cases selected}$$
$$R = \text{percentage of cases not selected (rejected)}$$

We do not have to know anything about the specific details of the formula to observe that by the Principle of Indifference, it must be true that

$$f(S, R) = D = f(R, S)$$

since the difficulty must be independent of the naming of cases. This functional formula says that by interchanging the independent variables, we must not change the value—unless the formula is wrong. By expressing this entire argument in functional notation, we simplify the rather cumbersome verbal arguments and numerical examples the previous chapter had to use to make the same point.

Functional notation can be used to indicate progressive increase in the scope of observations subsumed under a model. In our thermometer example in Chapter 3, the first model said

$$T = f(W)$$

where

$$T = \text{temperature shown on thermometer}$$

and

$$W = \text{water temperature}$$

The second model, which took into account the slow rise of the reading, said

$$T = f(W, I, t)$$

where

$$I = \text{initial air temperature}$$

and

$$t = \text{time since immersion}$$

(a small t is *almost always* used conventionally to indicate time). The final model was obtained from the view that

$$T = f(W, I, t, D)$$

where

$$D = \text{some variable characterizing the difference between}$$
$$\text{glass and mercury}$$

To further refine the model, we might have expressed D as a function of some other quantities, in a similar form. Even if we have not a clue as to *what* D depends on, we can use functional notation to express our *intention* to delve further into the matter—when we get around to it. We could write, for instance:

$$D = g(\ldots)$$

which makes our intention clear, and reminds us to return to this problem later on.

In the form

$$T = f(W, I, t, D)$$

we more or less assumed that the "independent variables" (the symbols in parentheses) were to be observed directly. Functional notation need not imply that the independent variables are "independent" in the sense that they do not depend on anything else. Instead, the independent variables may be interpreted as a declaration that, for the present discussion, we are not interested in dipping further into *their* functional dependencies.

Later, if we wish to emphasize these deeper levels of dependency, we can *compose* two or more functions, as in

$$T = f(W, I, t, g(\ldots))$$

whose meaning is quite clear and whose notation is quite compact.

As an example of composition of functions, consider our discussion of *dominance* of observers. If observer B dominated observer A, every observation A could make could be predicted from the corresponding observation of B. In other words:

$$A = f(B)$$

since nothing else but B is needed to determine A. The notation

$$B = g(A, \ldots)$$

or perhaps even

$$B = g(A, ?)$$

indicates graphically that while some information about B's observations can be derived from A's, A does not dominate B, since other factors are presumably involved.

If we have three observers where

$$A = f(B)$$

and

$$B = f(C)$$

then by composition we can write

$$A = f(g(C))$$

from which we can conclude that there is some other function, which we may call h, such that

$$A = h(C)$$

In other words, if A depends *only* on B and B depends *only* on C, then A may be said to depend *only* on C, even though we do not know the exact nature of any of the dependencies. Thus, we conclude that C also dominates A.

Because science "explains" by reducing one phenomenon to the terms of other phenomena, the notation of decomposition of functions is an appealing one. If we start with some function:

$$z = f(x, y)$$

and are not satisfied to stop with x and y, we can reduce them, in turn, to functions of other phenomena:

$$x = g(a, b, c)$$
$$y = h(c, d)$$

How, then, can a scientist ever commit a decomposition fallacy? There are two main answers to this question:

1. At some stage, he may have omitted something from one of the functional relationships, so that, for instance:

$$z = f(x, y)$$

was really

$$z = f(x, y, \dots)$$

Further attempts at decomposition will thus be in error, even though they might yield good approximate laws. We might call this the Fallacy of Incompleteness.

2. Even if the view is complete, the process of reduction must eventually stop, either because of the limited capacities of the observer—including limited patience—or because of the "reality" of the situation that simply will not admit of further reduction. Such a limit on the depth of reduction ultimately leads to a situation called "complementarity" of observation.

These two sources of fallacy will be our topics for the remainder of this chapter.

Incompleteness and Overcompleteness

The success of any physical investigation depends on the judicious selection of what is to be observed as of primary importance, combined with voluntary abstraction of the mind from those features which, however attractive they may appear, we are not sufficiently advanced in science to investigate with profit.

James C. Maxwell[8]

If we have omitted something from

$$T = f(a)$$

further decomposition cannot take place with logical assurance. This is the Fallacy of Incompleteness. What can it mean for

$$T = f(a)$$

to be incomplete? Obviously, the meaning has nothing to do with the specific equation relating a to T, for we have said nothing about equations, but only that T *depends on* a in some unspecified way. The functional relationship:

$$T = f(a)$$

could stand for the equation:

$$T = a$$

or for the equation:

$$T = a + 1$$

or for the equation:

$$T = \frac{2}{a + 1}$$

or for the equation:

$$T = 1 + a^2 - 3a^6 + 9^{-2a}$$

or, in fact, any one of an *infinite set* of equations involving a. By the use of functional notation, the members of this infinite set are lumped into a single element.

To speak of the functional relationship being "wrong" would have to mean the "true" equation is not in this set, which could happen in one of two ways. Either T does not depend on a—overcompleteness—or T depends on something in addition to a—incompleteness. What can be the basis of such a conclusion? Evidently, it can only be *observation* of the behavior of T and the behavior of a.

The reasoning goes like this. Suppose I observe a value of a and a value of T, such as

$$a = 7.5 \qquad T = 10$$

After making a number of other observations, I encounter

$$a = 7.5 \qquad T = 10$$

again. This much is clearly consistent with the idea that

$$T = f(a)$$

But if we then observe

$$a = 7.5 \qquad T = 25$$

something is wrong. Either T depends on something other than a that we are not observing, or we are measuring a or T incorrectly. Therefore, either we expand the scope of our observation so that

$$T = f(a, \ldots)$$

or we refine the grain of our measurement of a, or we discard the measurements as altogether erroneous. Which choice we make depends on how strongly we believe

$$T = f(a)$$

in the first place, and whether our science of observation is "sufficiently advanced" to make it possible to refine our measurements of a.

The other side of the coin is overcompleteness. T may not depend on a at all. We begin to suspect as much when we make a series of observations such as

$$
\begin{array}{ll}
a = 0 & T = 10 \\
a = 7.5 & T = 10 \\
a = -578 & T = 10 \\
a = 0.0003 & T = 10
\end{array}
$$

If, no matter how much we vary a, T does not change, we are wasting our time observing a.

Generally, of course, T will depend on *something*. If not, we simply will stop observing T, just as the fish stops observing the water. The problem is more complicated when, say, we imagine that

$$T = f(a, b, c)$$

and observe

$$
\begin{array}{llll}
a = 0 & b = 3 & c = 8 & T = 10 \\
a = 4 & b = 3 & c = 12 & T = 10 \\
a = -4 & b = 1 & c = 12 & T = 10
\end{array}
$$

Is T not varying because

$$T = f(b, c)$$

or is it because

$$T = f(a, b, c)$$

in such a way that certain effects "cancel out" over the range of our observations?

Which is the "right" answer? Which "explains" the data? For any finite set of *observations,* the set of *explanations* is infinite. For example, Figure 4.11 shows two explicit formulas for T that "explain" our three observations. One involves a and the other does not. Which is better? The observations we have cannot discriminate between them, so we may take our pick. The box is black. We cannot "see inside" to say which is the "true" structure, even to the extent of discriminating

$$T = f(a, b, c)$$

from

$$T = f(b, c)$$

The choice we make is up to us, and will depend on our capacities. If we can easily extend our scope of observation, but mental calculation is difficult for us, we may choose

$$T = f(a, b, c)$$

because that gives a "simpler" formula. On the other hand, if we have a good head for figures but not a very sharp eye for detail, we may choose

$$T = f(b, c)$$

so that we can observe less by thinking more. But as long as we are limited to this set of observations, we must refrain from saying that one view or the other is "right"—that is the basic rule of the black-box game.

			Model 1 $T = f(a, b, c)$ $T = \dfrac{(c - a)}{2} + 2b$	Model 2 $T = f(b, c)$ $T = (c - 10)^2 + 6(b - 2)^2$
a	b	c		
0	3	8	10	10
4	3	12	10	10
-4	1	12	10	10

Figure 4.11. Two possible "models" of the same observations.

Let us see how these arguments can be applied to the observations of the sequence of states of our music box. If we can designate by S the set of states that the superobserver can see, and by S_t (read "S-sub-t") the state he observes at time t, we can describe his observation of state determinedness by the functional relation:

$$S_{t+1} = f(S_t)$$

That is, the state at one instant of time $(t+1)$ is completely determined by the state at the previous instant t. We may read this relation out loud as

"S sub-t-plus-one equals f of S-sub-t"

or

"S-sub-t-plus-one depends only on S-sub-t"

This then, is the functional way of speaking about the property of being state determined.

But what about the other observers, the physicist and the inventor? If we designate their states by P and V, respectively, we observe that

$$P_{t+1} = g(P_t, \dots)$$

and

$$V_{t+1} = h(V_t, \dots)$$

because their views were *not* state determined. Take the inventor: To make his view determinate, given the behavior of Figure 4.5, he could expand his impression of what a "state" was to include the lights, or he could observe two successive states, as indicated by

$$V_{t+1} = h(V_t, V_{t-1})$$

The physicist, because he is tone deaf, is unable to refine his discrimination of states—unless he invents a tone-detecting instrument to give him finer grain. He, too, however, may observe past states and get determinate behavior, as indicated by

$$P_{t+1} = g(P_t, P_{t-1}, P_{t-2})$$

Because of our superobserver point of view, we can speak of the possibility of other observers broadening their scope or refining their grain or increasing their memory, but *they* do not have the information needed to make the choice.

Because the black box characterizes a situation in which we are already observing all we can observe, there is simply no basis within the observations themselves for choosing a *better* way for observing the box. The black box, through its behavior, can indicate that our view is *incomplete*, in the sense that it is not state determined. It cannot, however, tell us *how* to complete the point of view so it will become state determined. The observations are all we have to go on, and the inventor and the physicist are thus faced with an arbitrary choice among viewpoints, just as we might have chosen Model 1 or Model 2 of Figure 4.11—or any of an infinite number of other models that fit the data.

When we have two models that fit all observed data, we say that they are *isomorphs,* that is, of the "same shape." Mathematically, the two models would have to fit all *possible* data, but we use the term in a more limited sense—to fit all *observed* data. Given a particular level of observational knowledge, the best we can do on *logical* grounds is produce a *set* of models, all isomorphic, which fit those observations.

Black-box observation, once it no longer yields new observations, cannot resolve the isomorphism and select among the members of the set. Without opening the box, we do not know if it has gears inside, or electric circuits, or a trained monkey turning a crank.

But "opening the box" means decomposing one step further. At a *particular* level of observation, then, the choice of isomorph is strictly up to us. We may, in Figure 4.11, choose a model with

$$T = f(b, c)$$

or we may choose one with

$$T = f(a, b, c)$$

and, indeed, we may find one with

$$T = f(a, b)$$

or

$$T = f(a, c)$$

or even one with

$$T = f(a, b, c, d)$$

We may choose

$$T = f(a, c)$$

because b is hard for us to observe. The physicist may choose

$$P_{t+1} = g(P_t, P_{t-1}, P_{t-2})$$

because he is hard of hearing but has a good memory. But we also may choose

$$T = f(a, b, c)$$

because we do not notice that

$$T = f(a, c)$$

would suffice, or because we do notice but find the formulus unappealing, or because we are physicists and know that "physical systems don't behave that way," or because we are psychologists and know that "people don't behave that way," or perhaps because of a stubborn belief that "b is involved, somehow."

The arbitrariness of all these choices ensures that different observers will have a multitude of ways in which to interpret their observations, not just on the question of the precise isomorph to choose, but even on the question of "what is to be observed as of primary importance." And if we cannot even agree on this, on the functional form of the relationship, then we cannot be sure of correct reduction to other factors. Evidently, Maxwell's "judicious selections" and "voluntary abstractions of the mind" are subjects worthy of study unto themselves, and we shall take up that challenge in the succeeding chapters. For the moment, we must dispose of the second reason why decomposition might fail—complementarity of observations.

The Generalized Law of Complementarity

Within the scope of classical physics, all characteristic properties of a given object can in principle be ascertained by a single experimental arrangement, though in practice various arrangements are often inconvenient. . . . In quantum physics, however, evidence about atomic objects obtained by different experimental arrangements exhibits a novel kind of complementary relationship. . . .

Neils Bohr[9]

We have just seen that incompleteness leads to failure of the reduction strategy. Indeed, we could turn this idea around and take it as a *definition* of the intuitive idea of completeness. That is, no matter how we select among isomorphs, or decompose into more refined views, nothing essentially new will be found. We have also seen that the idea of completeness can only be an approximation—one that is based upon leaps of inductive faith and can thereby never be guaranteed.

We can accept failures of reductionism due to incompleteness because we have all experienced them in one form or another. We may have thought that

Course grade = f(exam grades, participation in discussion)

only to find out that

Course grade = f(exam grades, not disagreeing with the professor, sitting in the front row)

and then later to find out that

Course grade = f(exam grades, not disagreeing with the professor, sitting in the front row, previous reputation as a student)

There is, however, a second reason for failure of reduction, a reason that is more difficult for some of us to accept—the problem of complementarity. The physicists were the first to run up against this problem, which was fitting and predictable, since physics is further advanced in its ability to apply the reduction strategy. Other sciences or would-be sciences are generally so far from having *complete* views that they are not surprised that decomposition sometimes seems to fail. Had the idea of "complementarity" not arisen in physics—the very model of reductionism—it is likely that *nobody* would have accepted it.

The problem of complementarity arose on the subatomic scale, as physicists reached the last level of reduction they were capable of performing. So as to avoid involvement in the physical details, we shall explain the situation on a larger, more familiar, scale. Suppose we were taking a traffic safety survey to study how cars accelerated coming out of a toll booth. We would like to know each car's exact *position* and *velocity*.

Now suppose that we set up an automatic camera to photograph each car, after which we shall determine the position and velocity from the single observation, the photograph. There may be a problem reading the *exact* position of the car because the photograph of a moving car could be blurred. Therefore, we set the camera's shutter speed at its fastest position—to "stop the action" and get a nice clear picture from which the position can be determined quite precisely.

But how are we going to determine the velocity from a still photograph? We could observe how far out the foxtail on the aerial is extended, but not all cars have such rakish accessories. The most reliable method is to observe how blurred the photograph is, for the faster the car is moving, the more it will blur the film.

We can actually measure the length of the blur and divide it by the exposure time to get the velocity, which is the distance traveled in a particular interval. But notice the *complementary* nature of this method. In order to get an accurate velocity measurement, we must allow the blur to get long enough to measure reliably, yet, at the same time, in order to get an accurate position measurement, we must try to get the blur as short as possible. Therefore, *whatever* shutter speed we choose will involve some compromise, and a different observer, who sets his shutter speed differently, will see a somewhat different—or complementary—picture.

Our instinct is to escape this complementarity by refining our measurement, that is, by further reduction. We might, for example, get a less grainy film, so that finer and finer measurements could be made of shorter and shorter blurs. But suppose we have simply reached the limit of fine grain in film. Then there is no place to turn and we must content ourselves with the complementary views. In other words, *if* there is some limit to the grain of observation, *then* complementary views will result. *Whether* there is some limit to the grain of observation is, however, a different debate.

This debate has raged in physics for fifty years, and is only now being taken up in other fields. The principle as used in physics depends on the existence of indivisible quanta of energy. *If* there is a point where energy cannot be indefinitely divided, *then* there is, in our terms, a point below which the concept of decomposition of observation (or reduction) fails. To the physicist, there can be no act of observation without transfer of energy from the observed to the observer. Therefore:

... the notion of *complementarity* simply characterizes the answers we can receive by such inquiry, whenever the interaction between the measuring instruments and the objects forms an integral part of the phenomena.[10]

Put in this way, complementarity is a special case of failing to get a complete view. The observer would like to observe

$$z = f(x, y)$$

but because of the inherent grain of his observing, he can only achieve

$$z = f(x, y, \text{observer})$$

Though in general this will always be true, in many cases the amount of energy involved in the act of observation is so small that it can virtually be ignored. In our toll-booth experiment, light energy must go from the car to the photographic film, but the amount is so small that the inaccuracy in observation really would not bother the traffic commissioner. The light from the brightest flashbulb is not going to exert enough force to slow down or speed up the car, even if it does blind the driver and cause an accident.

In biology it is not so easy to avoid having the observer interact in unknown ways. To investigate life on other planets we must be sure that the rocket itself does not carry earth life forms or destroy the indigenous ones. Even a single one-celled organism carried on a rocket *could* have a devastating effect on a foreign ecosystem—a theme of countless science-fiction stories.

In the social sciences, too, the observer interacts in unknown ways with the observed. The anthropologist, because her very method is participant–observation, must have some effect on what she has come to observe. One wag quipped that the Zuni nuclear family consists of a father, a mother, two children, and an anthropologist.

Yet the *interaction* between observer and observed is too narrow a basis for a general concept of complementarity. By our Generalized Thermodynamic Law, we know that there may also be some mental reason for the same phenomenon to manifest itself. A classic case in anthropology is that of Robert Redfield[11] and Oscar Lewis[12] who observed the same Mexican village about a generation apart. The wide disparity in their views could hardly be accounted for by the changes that had taken place during that time, let alone the interference they made in the village life in the act of observation. From this and similar cases, we should have to conclude that two social scientists viewing the same scene are in much the same position as our inventor and physicist friend watching the music box.

The views of the inventor and our friend were, in fact, *complementary*. Though they were based on observation of the "same" situation, neither was reducible to that of the other. Nor were their views entirely independent, because certain things could be derived about each from the other. These are the essential ingredients in our idea of complementarity: two mutually irreducible points of view that are not entirely independent.

The physicist, however, thinks only of a more restricted type of complementarity. For him to speak of complementarity, there *must* be "interaction between the measuring instruments and the objects," because he believes that if there were no such interaction, further reduction would be possible; and not just *any* interaction produces complementarity—it must form "an integral part of the phenomena." By "integral," he means that the interaction cannot be avoided by any conceivable refinement of the experimental arrangement. If he cannot get finer-grained film, the physicist will construct a radar to observe the cars coming out of the toll booth.

To the physicist, the film technique was only one of "various arrangements" that he may use for "convenience." The physicist is no slacker—he will stop using the convenient experiment if it leads to complementarity that he can eliminate by using a less convenient arrangement. He will search for a film with a finer grain; he will get rid of the camera and build a radar; he will throw away all his radar equipment and build a laser. Never, but never, will he compromise with "convenience." Nothing less than "natural laws," an integral physical interaction, will force him to renounce his quest for the Holy Grail. No wonder such a man would hestitate to accept the idea of complementarity!

This view of complementarity we might call "absolute complementarity," because it depends on the idea that there is no alternative but to accept a *fundamental, integral* limitation on observation. The general systems view is based on a simpler premise, and therefore is more general. If, *for whatever reason,* observers do not make infinitely refined observations, then between any two points of view there will generally be complementarity. Since in almost every case there will be *some* reason for stopping short of infinitely refined observations, we can remove the condition and create the *General Law of Complementarity:*

Any two points of view are complementary.

The only exception we might have to make to this law is for the most careful observations of the physical scientists. But with they themselves admitting that the most careful observations *must* show complementarity, we need make no general exception.

We note, however, that observers do not always *care* if their views are complementary, which is a different matter. We are not all as fanatic as the ideal physicist. The economist and the sociologist looking at the same society naturally obtain different pictures, though there is at the same time some correspondence between them. Each view, if constructed with a modicum of care, will contain some information about what is "really out there," but they will never be completely reconcilable. Two economists, because they are competing for the same jobs, the same spaces in the journals, and the same political influence, will naturally be more concerned about the complementarity of their views. While an economist will attack another economist's writings as "false," he will simply ignore the sociologist's, for he *knows* they are talking about "different things."

It really matters very little whether or not one *might,* in principle, refine economic observations to the point where different economists would no longer have complementary views. We could muster a rather strong argument against it, but why bother? Economic data are certainly not sufficiently refined to eliminate complementarity.[13]

But supposing we *could* eliminate complementarity, would we want to? Just because we *seem* to be working towards an ideal does not mean we would like to achieve it. In playing superobserver, we may lose the joyous tinkling of the music box.

Reduction is but one approach to understanding, one among many. As soon as we stop trying to examine one tiny portion of the world more closely and apply some close observation to science itself, we find that reductionism is an ideal *never* achieved in practice. Reductionism is an article of scientific faith. It must be faith, for nobody has ever observed the final reduction of any set of observations. We may scoff all we like at the "nonscientific" idiots and their poetic explanations of the universe, but underneath, we know no more than they about why our methods work when they happen to work.

We have observed that reduction sometimes works, but to be objective we must admit that other methods sometimes work too. Because we are scientists, we believe that our methods will work more often, but there is no hard scientific evidence for that—only faith. Our methods work fairly often when we try them, but let us be honest about it. When they do not work on something, we soon stop trying.

Reduction works where reduction works. Where it does not work, we can keep trying, admit defeat, or pretend those situations "don't exist" or "aren't important." Let us face it, if we weren't such religious fanatics, we would have accepted complementary truth long ago. Perhaps we have been narrowed by the long struggle with established religion, but science itself how now become established, and may need its own reformation.

Thomas Blackburn[14] has written persuasively on the inability of scientists to accept the possibility of "truth" in complementary views— leading to what he calls an "underdimensioned" science. The conclusion of his essay can serve equally as a conclusion for our chapter on the humility with which we must interpret the black box of existence:

If the practice of science continues its present one-sided and underdimensioned course, new scientists will be recruited primarily from among those people to whom such a view of the world is most congenial. Yet such people are least fitted, by temperament and training, to hold in mind the complementary truths about nature that our looming task will require. Indeed, one may seriously question whether even an underdimensioned science can be maintained by scientists recruited from among those of lesser imagination, sympathy, and humanity. Neils Bohr's vision of the unity of human knowledge only echoes, a half-century later, that of Walt Whitman:

"I swear the earth shall surely be complete to him/or her who shall be complete. The earth remains jagged and broken to him or her who remains jagged and broken."

QUESTIONS FOR FURTHER RESEARCH

1. *Women's Liberation*

James Loy, in a review[15] of a book by Chance and Jolly, makes the following remark:

The book's main theme, that most catarrhine species are male-focal in their social structure, is not uncommon among students of the primates and probably stems from the fact that males of these species are typically much larger and more conspicuous than the females. Any action by an adult male immediately catches the eye of the observer. The question at hand, however, is whether or not the other group members as well as the observer orient to the male's behavior.

In our terms, these observers have been caught lumping states in too casual a way—if Loy is correct in his criticism. Human groups, too, are groups of primates in which the adult males are ordinarily larger than other members of the group—and may also be, like certain birds, more conspicuous for other reasons. Discuss the biases that may be introduced into the observation of human groups by this kind of lumping, and set down observational guidelines for overcoming such a bias in the observation of human beings and other animals in which there is sexual dimorphism.

Reference: Michael R. A. Chance and Clifford J. Jolly, *Social Groups of Monkeys, Apes and Men.* New York: Dutton, 1970.

2. *The Philosophy of Progress*

A soldier who was stationed on a small Pacific island during World War II took up shark hunting as a "hobby." When asked if he wasn't afraid to go after a shark with nothing but an iron bar sharpened into a spear, he explained, "What is there to fear? The shark is stupid and I am smart. They always respond the same way—they never surprise me."

We have strong intuitive notions of "higher" forms of life, always leading, of course, to man as the highest animal of all, and our society as the pinnacle of civilization. One of the functions of philosophy is to provide "proof" to back up our intuitive notions, and the philosophical branch of the general systems movement has begun to do just this. V. I.

Kremyanskiy, for example, has taken a point of view much like that of the shark hunter, which in effect says that one organism is "higher" than another if it possesses a wider range of properties—observations or behaviors it can make. Discuss Kremyanskiy's notion in terms of the lessons of this chapter.

> *Reference:* V. I. Kremyanskiy, "Certain Peculiarites of Organism as a 'System' from the Point of View of Physics, Cybernetics, and Biology." In *Modern Systems Research for the Behavioral Scientist,* Walter Buckley, Ed., pp. 76–80. Chicago: Aldine, 1968.

3. *Archaeology*

The reconstruction of the past through archaeology offers many opportunities for selective omission of information not conforming to present theory. At the highest level there is the very careful selection of the site itself. Within that site, the choice of exact places to dig out of all possible places may be dictated by theory, thus leaving undug regions as the basis for some later theory. And when digging a site it is the easiest thing in the world not to see certain things at all.

Binford gave an example of a much-dug site in southern Illinois that had failed to show remains of houses in earlier times simply because the investigators had been interested in those locations where the highest densities of broken pottery were found. We can speculate on how many houses a future archaeologist would discover if he excavated our cities in those places where he found the highest density of broken bottles.

Discuss how our society will look to various kinds of observers in the future.

> *Reference:* L. R. Binford, "Archaeology as Anthropology." *American Antiquity,* **28,** 217 (1962).

4. *Micrometeorology*

If a phenomenon is sufficiently rare and at the same time spectacular in our eyes, scientists have a way of shying away from its study. As a scientist, what would you think of reports of a phenomenon that had the following characteristics?

1. Floats through the air without visible means of support.

2. Glows with a reddish-purple luminosity, or sometimes bluish white, or someties golden yellow.

3. May kill instantly if touched, but passes through a window pane without breaking the glass.

4. Disappears in an explosion and blinding light, leaving only a smoke smelling of burnt powder and leaving a white spot on the floor as its only trace.

After reflecting a bit upon the matter, consult

> James Powell and David Finkelstein, "Ball Lightning." *American Scientist,* **58** 3 (May–June 1970).

Then discuss how it was that the very existence of ball lightning was denied for so many years, and then was acknowledged only as a rarity when compared with "ordinary" lightning.

5. *Physics*

Discuss the following seemingly derogatory remarks about physics:

> The laws of physics cannot distinguish between two males rabbits in a box and a male and female rabbit in a box.
>
> According to the laws of physics, a bumblebee cannot fly.

6. *Biology and Chemistry*

Since the molecular biologists have begun to experience success, the debate has raged over whether or not "all" of biology can be reduced to chemistry. W. M. Elsasser, in particular, has taken up the side of nonreductionism in biology, saying that the molecular biological and organismic views are complementary, based on the following argument:

> If we make elaborate measurements precise enough to determine the microsopic state of a system at a given instant, we can indeed find out what the state is but the disturbance engendered (for instance by the breaking of chemical bonds) would be so radical that the system would behave thereafter in a quite different way from the way it did before; it can no longer be considered as the same dynamical system. . . . We have killed the organism by our too detailed measurements.[16]

Discuss this argument and how it relates to the physics argument about complementarity.

> *Reference:* Walter M. Elsasser, *Atom and Organism.* Princeton, N. J.: Princeton University Press, 1966.

7. *Philosophy and Biology*

Discuss the following remarks, especially the use of the terms "real irreducibility," "logically untenable," and "methodological reasons":

> My general conclusion, then, is that, given the current state of biological science, there may be good heuristic reasons for not attempting in all possible areas to develop physicochemical explanations of biological phenomena, and good reasons for attempting to formulate specifically biological theories. This, however, is an argument which supports an irreducibility thesis for *methodological* reasons. Any attempts to twist this into a claim of *real* irreducibility for all time is, in the light of recent work in molecular biology, logically untenable, empirically unwarranted, and heuristically useless.

> *Reference:*   Kenneth F. Schaffner, "Antireductionism and Molecular Biology." *Science,* **157** (11 August 1967).

8. *Language Training*

One aspect of the disbelief in complementarity is the attitude of many scientists toward the learning of "foreign" languages. "Why take the time to learn Russian," they argue, "when any worthwhile results will be translated into English." In other words, there is nothing that can be expressed in Russian that cannot be expressed perfectly in English. Moreover, there is nothing to be learned about thinking from learning Russian, or any other language, for that matter.

Discuss the potential role of language learning in scientific education, and possible effects of the removal of all language requirements from scientific programs of study.

9. *Sky-watching*

Clouds have not changed their shapes since the Middle Ages, yet we no longer see in them either magical swords or miraculous crosses. The tail of the comet sighted by the great Amroise Paré was probably very little different from those which occasionally sweep across our skies. Yet he thought he saw in it a full suit of curious armor. Compliance with universal prejudice had bested the habitual accuracy of his gaze; and his testimony, like that of so many others, tells us not what he actually saw but what his age thought it natural to see.

Go out into a large open field—if you can still find one—lie on your back, and gaze up at the clouds for an hour or so. Make notes of the figures you see there, and later analyze those notes to see if you can detect the influences that have shaped your vision.

Reference: Marc Bloch, *The Historian's Craft,* pp. 106–107. New York: Vintage Books, 1953.

READINGS

RECOMMENDED

1. John R. Dixon and Alden H. Emery, Jr. "Semantics, Operationalism, and the Molecular-Statistical Model in Thermodynamics." *American Scientist,* **53,** 428 (1965).

2. Thomas R. Blackburn, "Sensuous-Intellectual Complementarity in Science." *Science,* **172,** 1003 (4 June 1971).

SUGGESTED

1. Lewis Carroll, *The Annotated Alice* (*Alice in Wonderland* and *Through the Looking Glass*), Martin Gardner, Ed. Cleveland: World, 1960 (also in paper).

2. R. L. Gregory, *Eye and Brain.* New York: McGraw-Hill, 1966.

NOTATIONAL EXERCISES

1. Show how the views of your friend the physicist were obtained in Figures 4.6 and 4.7.

2. Given the sequence
    O T T F F O T T F F O T T F F O ...
write down the three forms of shorthand:
   a. the ordered pairs
   b. the mapping
   c. the direct graph
Is it state determined? Is it determinate if we take pairs of states as our state?

3. Write down the first six states traversed by the system with the following directed graph of behavior—starting at state B:

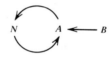

4. Suppose you are reading a book on farm economics and encounter the following statement:

   The amount of effort that seems to be devoted to loading every last bit of hay on the wagon seems to decrease when the travel time from hayfield to village decreases. On the other hand, older farmers seem to take more care than younger farmers, other things being equal—perhaps because transportation was slower in the old days.

Reduce this statement to a general functional description of the factors involved in the "amount of effort . . . devoted to loading. . . . "

5. Suppose you now read further in the book and find the statement:

   Travel time decreases as more modern tractors are used, as the quality of the road gets better, and as the hayfield is closer to the village. With modern roads, however, traffic enters into the picture, so that travel time definitely depends on the time of day.

Show how this refined view can be added to the functional model of Exercise 4.

6. Suppose you are reading a nontechnical description of a biological system and suddenly encounter the formula:

$$y = \frac{be^{-at}}{\sqrt{1 - b^2 e^{-ct^2}}}$$

You are about to put down the book in despair when you decide that it might be simpler to see what the author is talking about if you take a more general view, considering not the exact formula but simply the functional relationship that shows on what factors y depends. Extract such a functional description from the formula.

7. Suppose the set of states of a system is designated by S and that the state at time t is designated by S_t. How would you denote the seventeenth state? The fifth state after the state at time j? How would you indicate that the state of the system does not depend on the state previous, but only on the state previous to the previous state?

ANSWERS TO NOTATIONAL EXERCISES

1. First we obtain Figure 4.6 from Figure 4.1 by
   a. ignoring the G column, since the physicist is not looking at the green light
   b. grouping (1, 2), (3, 4), (5, 6) pairs because he cannot discriminate these tones
   c. naming the 6 sets of states S, T, U, V, W, X.
Then we use this table to translate the directed graphs of Figure 4.3, by replacing each of your state names with one of his. Thus, state a of yours translates into state S of his, and the cycle
$$a \quad n \quad i \quad k \quad a \quad n \quad i \quad k$$
translates into
$$S \quad T \quad U \quad W \quad S \quad T \quad U \quad W$$
and so forth for the other two cycles.

2. See Figure 4.12. Clearly the system is not state determined, since the mapping is not many to one but one to many in the case of T and F. It can be made determinate by taking pairs of states, according to the following mapping.

$$
\begin{array}{ccccc}
(O,T) & (T,T) & (T,F) & (F,F) & (F,O) \\
(T,T) & (T,F) & (F,F) & (F,O) & (O,T)
\end{array}
$$

which *is* many-to-one, and in fact, one-to-one.

3. $B \quad A \quad N \quad A \quad N \quad A \ldots$
Notice that this directed graph precisely characterizes a system that "knows how to spell 'banana' but doesn't know how to stop."

4. $e = f(t, a)$
where

e = effort devoted to loading every last bit of hay
t = travel time from hayfield to village
a = age of farmer

a. The order pairs

   (O, T)   (T, T)   (T, F)   (F, F)   (F, O)

b. The mapping

O	T	F
T	T? F?	F? O?

c. The directed graph

Figure 4.12. Solution to Notational Exercise 2.

5.
$$e = f(t, a)$$
$$t = t(m, r, d, T)$$

where

m = modernity of the tractor used
r = quality of road traveled
d = distance traveled
T = time of day

Also, we could write

$$e = f(t(m, r, d, T), a)$$

or possibly

$$e = f(a, m, r, d, T)$$

6. $y = f(a, b, c, e, t)$
However, e is a symbol conventionally used in mathematics for a certain constant (approximately 2.7), which enters into many formulas. If we recognize that e is, in this context, indeed the constant e, we can rewrite the functional relationship as

$$y = f(a, b, c, t)$$

since e is, in effect, only a mathematical symbol like the square-root sign, division bar, or minus sign. One of the things that confuses non-mathematicians is these standard symbols that might or might not be used in a standard way in a particular formula. You usually have to tell from the context whether e is the mathematical constant or a special symbol for some factor in the relationship. Since authors are not always careful to define their symbols, it may be difficult to make the decision in some cases.

7. The seventeenth state could be denoted by

$$S_{17}$$

while the fifth state after state j could be denoted by

$$S_{j+5}$$

The dependence on the penultimate state could be denoted by the functional relationship:

$$S_t = f(S_{t-2})$$

or

$$S_{t+2} = f(S_t)$$

5

Breaking Down Observations

Hazel's obsession with Hoosiers around the world was a textbook example of a false *karass,* of a seeming team that was meaningless in terms of the ways God gets things done, a textbook example of what Bokonon calls a *granfalloon.* Other examples of *granfalloons* are the Communist party, the Daughters of the American Revolution, the General Electric Company, the International Order of Odd Fellows—and any nation, anytime, anywhere.

As Bokonon invites us to sing along with him:

If you wish to study a *granfalloon,*
Just remove the skin of a toy balloon.

Kurt Vonnegut, Jr.[1]

It is related that Okubo Shibutsu, famous for painting bamboo, was requested to execute a *kakemono* representing a bamboo forest. Consenting, he painted with all his known skill a picture in which the entire bamboo grove was in red. The patron upon its receipt marveled at the extraordinary skill with which the painting had been executed, and repairing to the artist's residence, he said: "Master, I have come to thank you for the picture; but excuse me, you have painted the bamboo in red." "Well," cried the master, "in what color would you desire it?" "In black, of course," replied the patron. "And who," answered the artist, "ever saw a black leaved bamboo?"

Henry P. Bowie[2]

We settled ourselves about the round table at our drawing. I had only blue paint; nevertheless, I undertook to depict the hunt. After representing, in very lively style, a blue boy mounted on a blue horse, and some blue dogs, I was not quite sure whether I could paint a blue hare, and ran to Papa in his study to take advice on the matter. Papa was reading; and in answer to my question, "Are there any blue hares?" he said, without raising his head, "Yes, my dear, there are." I went back to the round table and painted a blue hare. . . .

Leo Tolstoy[3]

In this chapter we propose to discuss ways in which the limited mental powers of observers influence the observations they make. This difficult task is made doubly difficult by the psychological resistance that we encounter whenever we start to speak of the human being as limited in any way—but particularly in mental ways. Most people grudgingly accept that they are unable to flap their arms and fly, but intellectual people raise their hackles at the mere mention of limits to the intellect.

131

We have been subjected to so much bunkum about the "limitless capacity of the human mind" that as soon as we try to suggest "suppose there are certain limits to the human mind . . . " otherwise dispassionate readers turn purple. Yet the only capacity of my mind that seems limitless is precisely that capacity for fooling myself—particularly about my limitless mental capacity.

To say that mental capacity is limited is not to decree that there are some things we can never know. By accepting my mental capacity as limited, I willingly admit that one of those limitations is the inability to know my precise limitations. Yet this need not prevent me from exploring some of the consequences certain mental limitations would involve. I have every right to play the "if so" game with the idea of mental limitation.

To be specific, let us suppose that you are invited back by the mad inventor to see the latest model of his Edwardian music box, a special version for someone called Humpty Dumpty. He leaves you in the room to enjoy its tinkling, and you observe that it seems to have the same potential states as before:

$$S = (R, G, W)$$

which you have designated with the letters a through x, as in Figure 4.1.

Now let us pretend that the inventor has left you a bottle of "ten-star pair brandy," a few swigs of which have clouded your normally infinite mental powers. Had you not thus indulged, you would have recognized the 20-state cycle shown at the top of Figure 5.1.

The "ten star" on the brandy label is the strength of this insidious drink, and refers to its ability to limit your memory to precisely ten pairs of states. Under the influence of this seductive liqueur, you are not able to see the cycle at all, for you simply cannot remember enough states. Instead of a nicely state-determined behavior, you see only a drink-beclouded system, with every state coming at you as a complete surprise.

What can you do? You should not have imbibed, but regrets are too little and too late. The inventor will soon return, and he will think you are stupid if you cannot describe the behavior of his music box. Because you are desperate to defend the good name of superobservers, you decide to try *narrowing* your view, watching only the lights. You reason that since there are only two lights, there are only four states of the two-light system. Therefore, you are more likely to be able to hold all pairs in the sequence of lights in your memory.

Unlimited memory

Memory limited to 10 pairs

One instant	Next instant
(a, h)	(?, ?)
(h, o)	(h, o)
(o, v)	(o, v)
(v, e)	(v, e)
(e, g)	(e, g)
(g, n)	(g, n)
(n, u)	(n, u)
(u, d)	(u, d)
(d, k)	(d, k)
(k, m)	(k, m)
	(m, t)

Figure 5.1. Humpty Dumpty's music box seen with unlimited and limited memory.

What you see is shown in Figure 5.2, where only four pairs are needed to give a complete state-determined description of the behavior of the lights. Since this is well within your alcohol-limited capacity, you are able to see that the behavior of this smaller system—this "subsystem"—is state determined. Moreover, encouraged by your success, you try the same technique on the tones by themselves, and get the view shown in Figure 5.3.

What have you done? Under the inspiration—or the necessity—of the pair brandy, you have invented a new way of looking at the world. You have successfully *decomposed* Humpty Dumpty's music box into two parts, two *independent* parts, which are both state determined. But what good is such a decomposition? Before you had only one system, while now you have two. Isn't that making things more complicated?

If your brain had been a superbrain, unnumbed by drink, breaking the box into two subsystems would indeed have been more compli-

Lights		Substate name
R	G	
1	1	A
1	2	B
2	1	C
2	2	D

Graph	Pairs
	(A, B)
	(B, C)
	(C, D)
	(D, A)

Figure 5.2. The subsystem of lights.

cated. But then, if you had a superbrain, complication would not matter. Since you do not have a superbrain, decomposing the system like this has made your task appreciably easier. In fact, you can now hold all the necessary pairs in your ten-pair brain, with one pair left over for good measure.

The lesson is quite clear. If we have limited memories, *decomposing* a system into noninteracting parts may enable us to predict behavior better than we could without the decomposition. This is the method of science, which would be unnecessary were it not for our limited brains. The very existence of science is thus the best proof we have that human mental capacities are, in fact, limited.

Our little example of decomposition could be extended to illustrate the power of the method. Suppose the box had exhibited 180 states in one long cycle. If we could decompose it into a cycle of 20 and a cycle of 9, the number of pairs in memory would be reduced from 180 to 20 + 9 = 29. But we might go even further and reduce the cycle of 20 to a cycle of 5 and a cycle of 4, as we just did. Then the number of pairs needed would be only 5 + 4 + 9 = 18, or just 1/10 of the original number.

In general, in a system with one long cycle that has factors (such as 180 = 5 × 4 × 9), we may be able to reduce the number of pairs remembered to the *sum* of the factors: $5 + 4 + 9 = 18$. Consider a very long cycle of 10^{10} states. 10^{10} is the product of 10 tens, while the sum of 10 tens is only 100, or 10^2. Thus, if we can break down the system into ten independent cycles, we have made a reduction of $10^{10}/10^2 = 10^8$. While most of us can memorize a list of 100 pairs, how many could memorize ten billion?

Can such a decomposition into cycles always be done? At first sight, the answer would seem to be "no," but let us pursue the question a bit further. Suppose the inventor returns and you tell him that you have succeeded in understanding the behavior of Humpty Dumpty's music box. "It can be decomposed into two independent cycles," you announce proudly. But he looks unconvinced.

"Not so," he contradicts. "Humpty Dumpty specifically ordered a

Tone	Name
1	*V*
2	*W*
3	*X*
4	*Y*
5	*Z*

Graph

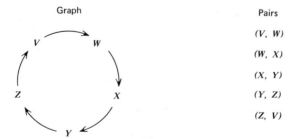

Pairs

(V, W)

(W, X)

(X, Y)

(Y, Z)

(Z, V)

Figure 5.3. The subsystem of tones.

machine with just one long cycle. Otherwise I would have sold him one of my standard machines, which *do* break down into a pair of independent cycles. Come next door into the warehouse and I'll show them to you. There are 144 different models. Quite impressive, if I do say so myself!"

You follow him next door and sure enough, there stand 144 music boxes all tooting and sparkling away. Your head is now clearer, and you can easily write down the cycles of three of them, as shown in Figure 5.4. But try as you may, you cannot get any of them to decompose the way Humpty Dumpty's did. (Try it and see for yourself.)

"Look here," you remark impatiently. "These cycles simply do not decompose. There must be something wrong with them."

"What do you mean? Of course they decompose. Just look at them!"

When you still look puzzled, the inventor tries to explain further. "Look! Look there! Just concentrate on the brilligance—that's the shorter cycle. Once you've seen that, you'll see the mimsy cycle immediately."

"What?"

"The brilligance. Keep your eye on the brilligance!"

"I don't know what you're talking about."

"I'm talking about plain old-fashioned brilligance. Don't play the fool with me."

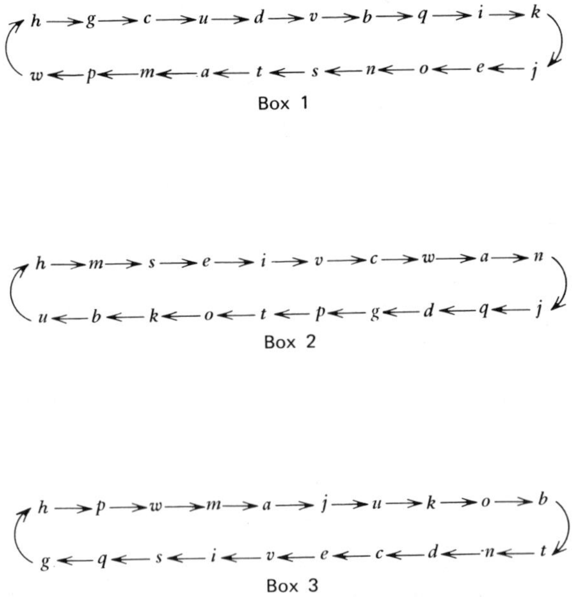

$$h \to g \to c \to u \to d \to v \to b \to q \to i \to k$$
$$w \leftarrow p \leftarrow m \leftarrow a \leftarrow t \leftarrow s \leftarrow n \leftarrow o \leftarrow e \leftarrow j$$

Box 1

$$h \to m \to s \to e \to i \to v \to c \to w \to a \to n$$
$$u \leftarrow b \leftarrow k \leftarrow o \leftarrow t \leftarrow p \leftarrow g \leftarrow d \leftarrow q \leftarrow j$$

Box 2

$$h \to p \to w \to m \to a \to j \to u \to k \to o \to b$$
$$g \leftarrow q \leftarrow s \leftarrow i \leftarrow v \leftarrow e \leftarrow c \leftarrow d \leftarrow n \leftarrow t$$

Box 3

Figure 5.4. The behavior of three music boxes.

"You don't have to shout. I don't know what you're talking about. I never heard of 'brilligance.' Could you explain it to me?"

"Explain? Explain? How do you explain brilligance? You might as well ask me to explain sound or light. You don't *explain* such things— you just point to them."

"Then would you mind terribly pointing it out to me?"

"Not at all. Look there at Box 1. See, the brilligance is *A*, then *B*, then *C*, then *D*, then back to *A* again. As I said, a cycle."

Hoping to clear up your confusion, you go to a blackboard nearby and write down your list of states, as you did in Figure 4.1. As he calls out the various values of "brilligance," you make a note of it next to the state name that you given. When you finally complete the list, he points out the five values of "mimsy" (*V, W, X, Y, Z*). Eventually, you get the list shown in Figure 5.5, giving the brilligance and mimsy for each state that the box displays. And, indeed, when you write down the

(brilligance, mimsy)

pair for each state in succession, you get the cycles shown in Figure 5.6—for all three of the boxes.

"Well," the inventor says. "Now do you see it?"

"I see that if you name your states that way, it comes out that way. That's what I see. But I still can't figure out what you call 'brilligance' and 'mimsy.' They're not real qualities at all. They're just figments of your imagination—like blue hares or red bamboo groves."

"And I suppose, Mr. Superobserver, that your imagination has no figments?"

"Of course not. I see only *real* things—like that red light going on and off."

"I never heard of 'red light.' Could you explain it to me?"

"Explain? Explain? How do you *explain* a red light? You might as well ask me to explain. . . . " Suddenly you stop, and turning contritely toward the inventor, all you can mutter is a feeble "Oh yes I see."

What exactly do you see? You still do not "see" brilligance or mimsy, but you now understand that the inventor is not kidding when he says that *he* sees them. While this realization is difficult for you to accept, you know that

. . . each one of us, in the course of development, has painfully worked out a set of assumptions as to what is real and what is important in determining our behavior; secondly, that these assumptions give meaning to our lives and offer some protection from fear and uncertainty; and, thirdly, that even personal attempts to modify such deeply-rooted assumptions arouse anxiety and resistance which can only be overcome by serious psychological effort.

Elliott Jaques[4]

For most ordinary experience, the concepts of "red light" or "musical tone" are adequate for the observation of systems. In this warehouse, however, if you can teach yourself to "see" brilligance and mimsy, your view of the world will be simplified. Though they may seem artificial at first, there is no doubt at all that given enough exposure to these boxes, you would come to recognize "brilligance" and "mimsy" as readily as you recognize "red light" or "musical tone."

In the same way, the physicist comes to recognize "entropy" and "density"; the chemist, "valence" and "PH"; the electrical engineer, "carrier frequency" and "impedence"; or the economist "profit" and

R	G	W	Name	Brilligance	Mimsy
1	1	1	a	A	W
1	1	2	b	C	W
1	1	3	c	C	X
1	1	4	d	A	Z
1	1	5	e	D	W
1	1	6	f	?	?
1	2	1	g	B	W
1	2	2	h	A	V
1	2	3	i	A	Y
1	2	4	j	C	V
1	2	5	k	B	Z
1	2	6	l	?	?
2	1	1	m	B	X
2	1	2	n	B	Y
2	1	3	o	A	X
2	1	4	p	C	Y
2	1	5	q	D	X
2	1	6	r	?	?
2	2	1	s	C	Z
2	2	2	t	D	V
2	2	3	u	D	Y
2	2	4	v	B	V
2	2	5	w	D	Z
2	2	6	x	?	?

Figure 5.5. How brillig is each separate state? How mimsy does it calculate?

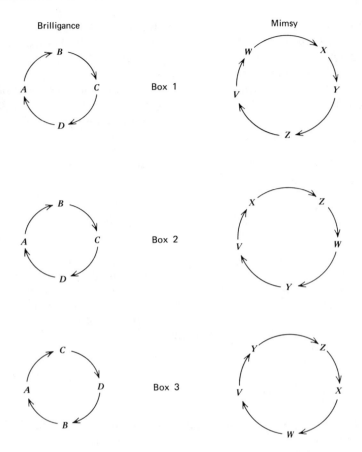

Figure 5.6. Brilligance and mimsy cycles for three music boxes.

"marginal utility." Your resistance to mastery of this new point of view is precisely the same as the resistance students display to learning that "mass" and "weight" are not the same thing. If you practice long enough and hard enough, you will soon see that the inventor is not mad at all, but as clever as your physics instructor, who knew those difficult things all along.

What is the lesson in our little tale? Recall the Principle of Indifference:

Laws should not depend on a particular choice of symbols.

Fix your attention on the word "should." Now notice that the difference between your (R, G, W) and the inventor's (brilligance, mimsy) is strictly one of choice of symbols, since you both have exactly the same power to resolve states. But because of your built-in apparatus for partitioning the world, you cannot reduce his 144 music boxes to simple machines with two independent cycles. Eventually, you may learn how. But in the meantime, if you get drunk again, you will not be able to see lawful behavior in any of the 144 boxes. Therefore, we may conclude this section—and introduce our topic for the chapter—by stating the *Principle of Difference:*

Laws should not depend on a particular choice of symbols, but they usually do.

The Metaphors of Science

We proceed in step-by-step discussion from inference to inference, whereas He conceives through mere intuition. Thus, in order to gain insight into some properties of the circle, of which it possesses infinitely many, we begin with one of the simplest; we take it for a definition and proceed from it by means of inferences to a second property, from this to a third, hence a fourth, and so on. The divine intellect, on the other hand, grasps the essence of a circle *senza temporaneo discorso* (without the use of profane reasoning) and thus apprehends the infinite array of its properties.

Galileo[5]

Let us review where we stand in our discussion of method. Trying to cope with unfamiliar, complex phenomena, we try to

1. get a "complete" view—one broad enough to encompass all phenomena of interest—so we are not surprised;
2. get a "minimal" view—one that lumps together states that are unnecessarily discriminated—so we do not overtax our observational powers;
3. get an "independent" view—one that decomposes observed states into noninteracting qualities—so as to reduce the mental effort required.

While these goals can often be met, the resulting way of looking at the world may not be "satisfying." That is, it may not conform to the psychological categories we have either inherited or learned from the past. Once again, our limited powers are the essential reason why we want "natural" or "satisfying" points of view, because we shall not be able to carry a different point of view in our heads for every moment of our lives.

In other words, we are like the handyman who can only carry a single box of tools to each job, but has to be ready to do plumbing, electrical work, painting, carpentry, glazing, metal work, or what have you. From time to time, the handyman will eliminate one tool in favor of another that he finds to be more generally useful. In doing so, he makes the assumption that the kinds of calls he is going to get in the future are somewhat like the calls he has been receiving in the past.

How does the handyman know that future calls will be like past calls? It is only an article of faith—one that we have encountered before. Perhaps we should give it a name—*the Axiom of Experience:*

The future will be like the past, because, in the past, the future was like the past.

Underneath it all is Patrick Henry's observation:

I have but one lamp by which my feet are guided, and that is the lamp of experience. I know of no way of judging of the future but by the past. . . .

In other words, what else can we do? By studying the past, the handyman may be able to develop a more useful box of tools, and so it is for the scientist.

But the Axiom of Experience can, like all of our principles, be turned around, to become a definition of what we mean by the word "like":

Two things are alike if one in the present can be substituted for one in the past.

To say that the future will be like the past means that certain properties that we regard as important will be preserved. But *which* properties?

Through the study of poetry, we apprehend the magnificent profusion of ways in which one thing can be "like" another. The very essence of poetry is the *metaphor*—literally, that which "transfers over." Metaphor talks of one thing in terms of another, as in:

My love is like a red, red rose . . .

or

I embraced the summer dawn . . .
(*J'ai embrassé l'aube d'été.*)

A metaphor works only because we know or feel we know, some properties of the one thing that we can transfer over to the other. We do not know how Burns felt about his love, but we know something about how *we* feel in the presence of a red, red rose. In making his comparison, or metaphor, Burns depends on the universal experience of roses and color perception. If we had no experience of roses, the analogy would have no more meaning than to say "My love is like a slithey, slithey tove."

One of the problems induced by specialization in the sciences is that scientists in different fields have few common experiences to serve as the basis for communication. Englishmen have their gardens, so even if they are celibates they understand something of what Burns was saying. Frenchmen understand what it means to "embrace," so even if they sleep until noon all summer, they understand Rimbaud. And, even though we may never have known love or dawn, we can use Burns and Rimbaud to transfer some of our understanding of roses to an understanding of dreams of a summer morning.

Let us put the last paragraph into nonpoetic terms, using a metaphor of our own. To say that something is "like" something else means that the image of the one can be made to "depend on" the image of the other. Therefore, *metaphor is like function.* Instead of

My love is like a red, red rose . . .

we could write

$$\text{loved one} = f(\text{rose}, \dots)$$

which states that loved one is like a rose, in some unspecified way f. Or, instead of

I embraced the summer dawn.

which likens the dawn to a loved one, we could write

$$\text{summer dawn} = g(\text{loved one}, \dots)$$

At this level, science and poetry are very much alike. The poet starts with a metaphor and then may elaborate the details of how his love is like a rose, or how the dawn is like an embraceable goddess. The scientist starts with his complete view and refines and simplifies, ultimately reducing the original function to a function of other things. Like the poet's, his ultimate reductions are assumed known and therefore left undefined.

In science as in poetry, the essential quality is not the finished metaphor itself, but the *process of transformation,* this is, the *process* of making the metaphor. And as the fabric of poetry or science is made, metaphors may be built upon metaphors, functions upon functions. If

$$\text{loved one} = f(\text{rose}, \dots)$$

and

$$\text{summer dawn} = g(\text{loved one}, \dots)$$

then there is a sense in which

$$\text{summer dawn} = g(f(\text{rose}, \dots), \dots) = h(\text{rose}, \dots)$$

or more poetically,

The bud of night, tightly drawn, spread its petals into dawn.

Poetry based on other poetry is often called "academic" poetry, because its basis of reference is not direct experience of the world, but experience of other poetry. Science based on other science is often called "academic" science, for the same reasons. Both, in the extreme, are forms of mathematics—a kind of Glass Bead Game[6] played high above the plane of profane life.

When Russell said that mathematics was completely without content, he was flattering the mathematicians, not condemning them. In their image, the ideal of mathematics is to erase all traces of profane origin. Although this ideal may motivate mathematicians, it cannot be reached, any more than Tarzan could learn to read and speak English from books he found in the jungle. We would do well to remember Tarzan as we study the ways in which scientists use their metaphors, the ways in which they transform knowledge from one situation to another.

In science as in poetry, the meanings of the words we use must ultimately be rooted in observation. "We proceed in step-by-step discussion from inference to inference," but we must begin with some property of the circle. In the same way, we may proceed to an understanding of the dawn in step-by-step metaphor through Burns by way of Rimbaud, but we must begin with the meaning of "rose."

We can no more eliminate profane material than the Magister Ludi could purify his Bead Game. In the end, as Hesse's hero discovered, there is always profane material because we *are* profane material. We are the distillation of millions of generations of profane material, most of which lies so deep within us we imagine it to be outside, in the "real" world. It is no accident that our eyes are sensitive to the range of radiation that maximally penetrates our atmosphere from the sun. Animals in closed caves are blind—blind animals probably did not evolve in the light of day. By examining vision, we can learn about the world of the past in which that vision evolved; and by examining the metaphors of science, we can learn about the limitations of the brains that do science. In short, we can learn about ourselves, which is really why all of us are playing this incredible game, call it poetry, beads, or, if you will, science.

Boundaries and Things

> The white chrysanthemum
> Is disguised by the first frost.
> If I wanted to pick one
> I could find it only by chance.
>   Oshikochi No Mitsune[7]

One of the most deeply buried metaphors of science is the concept of a "thing" or "part" that can be separated cleanly from other things or parts. The metaphor is so deep that we seldom know when we are using it. The anthropologist speaks of the "social organization" of a tribe, as if it were a box of matches he could carry around in his pocket. But when the fledgling fieldworker arrives at his site, "social organization" is nowhere to be seen. The economist speaks of the "gross national product" as if it were a hog being fattened for slaughter. The mighty fall from power as the GNP falls in weight. But where would we go to have a look at the gross national product? To the Treasury?

These "things" or "parts" are the possessors of "properties" or "qualities" that they carry around with them as a box carries matches or a hog carries fat. These properties can be isolated from other properties by isolating the "thing" from other "things." To weigh the hog, we must remove it from its pen, clean it up, and put it on the scale. To weigh the GNP, we create a special statistical arm of the government, employing hundreds of economists to isolate it from all the other figures with which it wallows.[8]

Our use of the "part" or "thing" metaphor is closely allied to our experience of physical space, and particularly to our experience of "boundaries." As Leonardo observed, "the boundary of one thing is the beginning of another."

On the surface of the earth, we can draw a line around something and immediately discriminate "inside" from "outside." Even if the line is as tortured as the one in Figure 5.7, we can carefully determine whether any point, such as *P*, is inside or outside. "Outside" is that very far away place that corresponds to our idea that things placed at a great distance do not affect each other in appreciable ways. Thus, we can determine if *P* is outside by starting at *P* and heading toward the *known* outside, far, far away. If we count the number of times we cross the boundary, when we finally get to a point that is definitely outside we can calculate whether *P* was inside or outside, as shown in the Figure.

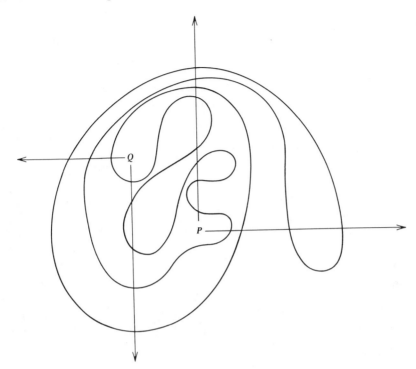

Figure 5.7. How to tell the inside from the outside. Starting at point P, we draw a line out to the far reaches, which we are sure is outside. Since the line crosses the boundary an *even* number of times, P is "outside." Q, on the other hand, always gives an odd number of crossings, and so must be "inside." "Inside" and "outside" are thus well-defined concepts for the finite bounded figure.

   The idea of clean separation of one part from another is so deeply ingrained that we have confidence that we can always separate inside from outside, even though it may take much mental effort. By analogy, we apply this concept to all our systems, using the term "system" to mean the "inside" and "environment" to mean "outside." By the Principle of Indifference, we might imagine that we can call either one "the system," for one man's system may be another man's environment. By the Principle of Difference, however, we suggest the choice may be very important to our view of the world.

   Not all of our systems exist in the physical world, so the idea of boundary is generally only a figure of speech. Even so, we already encounter difficulties of reasoning when we are dealing with systems that have boundaries in physical space. Problems particularly arise because we are influenced in our choice of boundary by our past experiences, or the experiences of our ancestors. Because these influences are excellent guides most of the time, they are all the more bothersome to clear out of our minds when they do not work.

For example, when we choose boundaries, we are powerfully influenced by easily recognized physical features. A place where sharp color change is seen, where difference in texture is felt, where solid meets liquid or liquid meets gas—all these and many more make popular boundary choices. On the other hand, we hesitate to define a boundary between two solid bodies rigidly attached so that they always move together. If we look for a white chrysanthemum in the frost, we find it only by chance.

Most of the time, we think of the boundary of a person as his skin, for there we find a solid–gas interface with a distinct color change. The lungs are a bit of a problem, for they are shaped much like the tortured curve of Figure 5.7, but by closing our mouth we make the air inside our lungs a definite part of our body. On the other hand, there is a less defined quantity of air that moves about with us even though it is not in our lungs. There is no definite boundary beyond which the envelope of air stops, but as we move, some air moves, and when we stop, it stops. Some of this air is circulated by the force of our breathing, but not enough to accomplish the important job of moving stale air away from our body and bringing fresh (oxygenated) air toward it.

A mechanism exists to accomplish this essential change of air, but few of us have ever given it much thought. Convection currents are established because the stale air is slightly warmed before it leaves the body. Warm air weighs less than the cooler, unused air, and so tends to rise away from the mouth and nose and be replaced. Space capsules have to be designed to accomodate the bodily needs of an astronaut under abnormal conditions. Under zero gravity, convection currents—which depend on the weight differential between warm and cool air—do not operate. Specific provision must be made for circulating the air that forms the outer part, from one point of view, of the spaceman's body. Without explicitly recognizing that this air was part of the "spaceman system," we might have designed a capsule in which an astronaut would suffocate in his own breath.

A second example also concerns the human body. Recently, physical anthropologists have engaged in active controversy over why human beings have very little body hair as compared with their fellow primates. One school argues that with less hair man was able to dissipate heat more effectively than other primates and thus hunt in the middle of the tropical day. Another school claims that the hair was a breeding ground for parasites. A third claims man lost his hair during an aquatic phase. Yet these and other theories all fail to explain why any hair, such as on the head, was retained.

We commonly consider hair to be part of the body, because it is attached to it. When we are thinking of thermal problems, parasite problems, or swimming problems, it is useful to consider the hair in such a way. But this accepted mode of thought blinds us to the possibility that, for some purposes, hair is better thought of as being *outside* of the body. Unlike the material in body cells, material once secreted into hair no longer participates in the body's physiological processes. Since material in the hair was *once inside* the body's physiological system and is *now outside*, it is useful for the physiologist to think of hair as excrement—just like perspiration, urine, feces, and, for that matter, toenails. It seems contrived to think of hair in this way, not only because the hair is attached but because the rate of excretion seems so slow. Yet the concept of hair as part of the environment makes the physiologist think of examining hair to see what it carries away from the body. As it happens, certain trace elements are carried away most effectively in the hair; which may partially explain why *all* body hair could not be eliminated.

These examples could be multiplied endlessly to demonstrate how our built-in notions of "natural" boundaries may make a difference in the effectiveness of our thought. Still, our ancestors gave us a pretty good set of tools for dividing system from environment in physical space, tools that we should not simply ignore. When, however, we begin to encounter systems without well-defined physical boundaries, the boundary metaphor lures us more easily into attractive, but false, reasoning.

The trouble begins because even physical boundaries are not quite what we imagine them to be. A boundary may quite intentionally go "through" something that we can recognize as a "part" in another context—like the saloon with the bar in Texas and the floorshow in Oklahoma, to circumvent the liquor laws in one state and the public decency laws in the other. The problem here is that a "boundary" may not be infinitely thin, precisely so it *can* partake of both system and environment. Rather than separating, such a boundary *connects*.

In order to make it perfectly clear that we are not talking about a perfectly thin, perfectly separating line or surface, systems thinkers use the term "interface" to describe that part of the world that, like the two-faced god Janus, looks both inside and outside at the same time. "Interface" is a more useful word than "boundary," for it reminds us to pay attention to the *connection,* and not just the separation, between system and environment.

The boundary metaphor permeates systems thinking more through *diagrams* than through words. Conventionally, a "part" is represented on paper as a region surrounded by a boundary—a rectangle or circle or any other simple closed figure. A "connection" is represented by a line or an arrow. In Figure 5.8, we see several archetypical diagrams of the division into system and environment, with the interface specifically included. Note that, for the systems thinker, the diagrams are supposed to be equivalent, since the shape and size of boxes, the length and curvature of the lines, or the placement of the boxes and lines on the diagram do not enter into the "abstract" structure being described. Still, we are not altogether indifferent to these things, and sometimes a well-drawn diagram can perform wonders for our understanding.

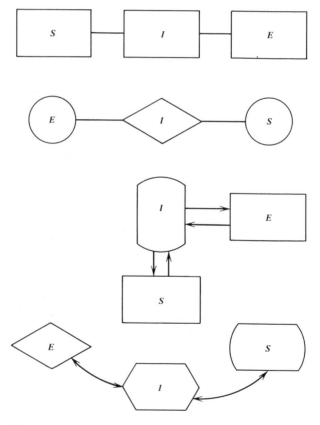

Figure 5.8. Various equivalent abstract representations of the system–interface–environment division: S = System; E = Environment; I = Interface.

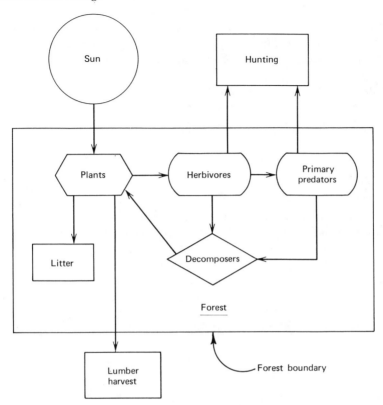

Figure 5.9. A simple conceptual model of a forest consisting of "parts."

But, once again, the very power of such diagrams may lead us down the primrose path. If the shape or placement of the boxes can suggest, they can by the same token mislead. Even more important, though, is the translation of diverse systems by the boundary metaphor. Not all systems, by any means, can be separated from their environments in the sharp, clean way suggested by such a diagram as Figure 5.9. In the first place, the boundary around a "forest" is never quite as clear as we first imagine. Even more misleading is the idealized "boundary" between such "parts" as plants and herbivores. Although we might possibly draw such a boundary in a mathematical sense, it would be so involuted as to make Figure 5.7 look like a perfect circle in comparison. Besides, animals move about, so can we really draw a meaningful boundary between plants and animals, or between herbivores and predators—which might, moreover, be one and the same animal? We cannot, any more than we can separate labor from management in the General Electric Company, vitamins from minerals in a watercress sandwich, or brilligance from mimsy in a silly music box.

We shall often employ directed graphs with bounded boxes as structure diagrams to aid in systems thinking. But when we say implicitly in these diagrams that

My system has a sharp, sharp boundary

we are really saying the same kind of thing as

My love is like a red, red rose.

Therefore, if we as scientists set out to make more specific conclusions about a system, we shall have to progress from the rich metaphor of the poet to a more precise description of separation. Moreover, such structure diagrams are limited—or, rather, we are limited—to perhaps 15 boxes at a glance. Beyond that, our minds begin to boggle, and we need further support, which will come from several directions.

Qualities and the Principle of Invariance

There are two kinds of people in the world—those who divide everything in the world into two kinds of things and those who don't.

Kenneth Boulding

What do we mean by a quality? As the inventor taught us, we cannot *explain* what we mean, except by pointing to the states which have different values of this "quality." We call such a definition by pointing "ostensive definition." Though we might explain one set of qualities in terms of another, we are simply concealing the fact that the primordial set was obtained by ostensive definition. True, we are very far removed from such primordial definition, so far that we do not discriminate between original and derived qualities. For now, however, we would like to trace things back a bit closer to the first qualities—the rose that is a rose that is a rose.

Qualities have a mental function for observers with limited memory. We may think that certain qualities are more "natural" than others, but that means merely that we are more accustomed to observing in those terms. By "accustomed," we may include "genetically accustomed," that is, our ancestors found certain qualities more useful than others and over many generations evolved abilities to "see" those qualities more readily than others. These abilities have been passed down to us, which is why we believe that "redness" is a more "real" quality than "brilligance."

And, of course, it may be more real, in the sense of being found useful, on the average, more frequently than "brilligance." But we are not concerned with averages here. In particular situations, *any* quality might be the best one for simplifying our view. "Red light on" may be just the thing we need when driving through traffic or using an electric stove, but it may be useless in walking through the jungle or using a mad music box.

As we work in less and less familiar situations, our inherited and learned perceptual capacities become less and less effective. We do not often have freezing and flowers at the same season, so we are relatively unequipped to find a white chrysanthemum in the first frost. Had we inherited infrared vision from our ancestors, we would have no trouble at all. In the crazy world of Humpty Dumpty we find ourselves without any resources at all.

If we leave the question of inheritance of quality perception and try to treat the question abstractly, we may make the following analysis. The inventor is speaking of a quality which he calls "brilligance." To try to understand what he is talking about, we first ask him to point out which states have which *quantities* of brilligance. If all states have the *same* quantity of brilligance, then the *quality* of brilligance is of no use in decomposing the behavior: it is simply synonymous with "being a state," or "existing."

The "quantities" of brilligance need not be numerical quantities. In the case of brilligance, there were four possible quantities:

$$(A, B, C, D)$$

To be a quality, however, brilligance must have *some* quantity associated with *each* state observed, but this can be handled, if necessary, by adding a quantity to the list called "don't know" or "doesn't have." This value may also be associated with states not yet observed, as we did with the question mark for states (f, l, r, x) in Figure 5.5.

Although we could guess what value of brilligance unobserved states would have if they *were* observed, we could only do that on the basis of some other qualities, such as (R, G, W). In fact, we wrote down (f, l, r, x) only because we had seen these states before, or inferred their existence from previous knowledge, using the Cartesian product. We could as well have eliminated these states from consideration had we never observed other music boxes. But even after a few minutes of observation, we begin to transfer what we learn in one situation to what seems to us to be somehow "alike." One of the advantages of decomposing a system into qualities is the possibility of extending our view to states not yet observed. If we merely observed 20 undifferentiated states in a cycle, we would have no basis on which to describe other states. The 20 states simply have the quality of existence, nothing else.

In our original observation of the black box, however, we *started,* for some reason not explained, by recognizing that there were three "parts," or qualities, (R, G, W). By observing the separate grain of each quality, we were able, by the Cartesian product, to extrapolate to possible states not yet observed. This extrapolation may be wrong in any particular case, but it is the only plausible way to go beyond the immediately given data.

In the inventor's view of the Humpty Dumpty music boxes, there are only $5 \times 4 = 20$ possible states. Since all of them are seen in the behavior, he cannot surmise the existence of other possible states as we can. Whether or not his view is better than ours depends on whether or not these music boxes can ever get into states (f, l, r, x). If, for example, he is using a new type of whistle that does not have a sixth tone, our extra states will prove to be extra mental baggage.

In sum, a *quality is a way of grouping* the states of a system. For example, the quality of *mass* is ultimately defined by displaying states of a system in which masses are the same or different. If we want to *measure* mass, we must introduce another operation besides "sameness" and "difference" for states, something like "greater than." But for now we are only concerned with the simple act of division which indentifies the quality in question.

Scientists sometimes speak of two kinds of qualities—*extensive* and *intensive*—according to what happens to the quality when the system is divided into parts. If we break a chocolate bar in two pieces, each piece has a different mass than the original: thus mass would be called an *extensive* quality, since it depends on maintaining the full extent of the system. On the other hand, when we break the cholcolate bar in half, each piece retains the same "chocolateness"; which is therefore said to be an *intensive* quality. Or, to take a more physical example, each half has the same density, so density is said to be an intensive quality.

Some systems researchers have tried to take over this distinction from physical sciences, but in doing so they make a fundamental error in their understanding of the idea of "quality." The confusion is once again a confusion of relative and absolute thinking, because intensive and extensive are defined *relative to some act of breaking.* Take density, for example. If we cut a chocolate bar, the density of each of the two halves will be the same as that of the original. If, however, we mentally decompose the chocolate bar into the qualities of flavor and consistency, neither "part" (not "flavor" nor "consistency") has any density at all. Therefore, density is an intensive quality with respect to cutting with a knife, but something else with respect to a certain mental cleavage.

A less technical example may help even the physicists. In our everyday English speech, we readily distinguish in forming plurals between intensive words such as "milk" and extensive words such as "watermelon." If I add milk to milk, I get milk; but if I add a watermelon to a watermelon, I get watermelons. In reverse, if I divide a glass of milk I get two glasses of milk, but if I divide a watermelon, I simply get two watermelon halves. Where does the distinction lie?

Sugar is like milk, and so is popcorn. The size of the pieces seems to have something to do with it, for they determine the customary way we divide the substance. I will never forget the hectic Saturday at the dried foods counter when a little old lady came up and demanded, "How *many* cranberries in a pound?" Cranberries are on the borderline between intensive and extensive—she could have asked the question about walnuts, but not about peanuts. That is, we may *count* walnuts if we want to, but we find it more sensible to *measure* peanuts. Yet the feelings we have about these plurals derives only from our customary way of breaking. Avogadro made scientific history by daring to ask, "How many atoms in a pound?"

The definitions of intensive and extensive properties may be turned around and taken as a definition of "breaking." The physicist gives me a list of intensive and extensive properties and then says:

> If the intensive properties remain the same, then you have properly broken the system.

In other words, if I pick the nuts out of the chocolate bar and make a pile of nuts and a pile of chocolate, the two piles will have different densities. We can either conclude that density is not an intensive property, or that I have not followed the physicist's rule for breaking.

For almost every "thing" in our normal experience, we carry in our minds a set of rules that are "proper" for breaking it. If you and I were to divide a bag of peanuts, you would laugh at the suggestion that we each take half of each peanut, or even that we count out equal numbers of peanuts. We are confused about cranberries partly because they are less familiar than peanuts, and partly because they are near the fuzzy border—the capacity of one hand—that separates one kind of proper breaking from another.

Instead of focusing on the properties, then, we could focus on the *types of breaking,* keeping one property or set of properties *invariant.* Then we may say that there are two kinds of breaking, one that preserves this property and one that does not. Breaking into parts, however, is only one of the scientist's metaphors, or "transformations." We can apply the same idea to any other transformation and derive the *Invariance Principle:*

With respect to any given property, there are those transformations that preserve it and those that do not preserve it.

Moreover, we may turn the emphasis around and restate the Invariance Principle in terms of the transformations being fixed:

With respect to a given transformation, there are those properties that are preserved by it and those that are not.

In general, then, a property or quality may be characterized by the transformations that preserve it or a transformation may be characterized by the properties it preserves. We value certain transformations as tools because they preserve properties in which we are interested. When we say that the four diagrams in Figure 5.8 are "the same," we are saying that we may apply changes in scale, shape, and orientation without changing the properties we value in this instance. Therefore, the properties that the permitted transformation does change are the "unimportant" properties.

Conversely, we say which properties are important by prohibiting the corresponding transformations. We must not erase lines, add boxes, or reverse arrows. In particular cases, we may make one of these transformations without changing the meaning—like reversing all the arrows in Figure 5.8. We are not talking about particular cases, however, but sets of cases. In this instance, we are talking about the property of "being a structure diagram of a certain structure," which property is generally not preserved by reversing all the arrows.

In general, we cannot say precisely what we mean by a property because there are an infinite number of possible transformations. If we change the color of the boxes, do we preserve the property of being a structure diagram of a certain structure? The answer is yes, but we do not know that from the four examples of Figure 5.8. In the same way, we could not be sure from that limited set of examples whether or not it was permissible to reverse all the arrows.

We may summarize all this by restating the Principle of Invariance in the following way:

We understand change only by observing what remains invariant, and permanence only by what is transformed.

Partitions

Now we have always recognized that the operation of isolation is one which is almost indispensable in the common sense handling of the world around us, but at the same time it is an operation which is never sharply performable, for everything has a fringe of involvement with its surroundings.

P. W. Bridgman[9]

As an example of the division of a system into nonbounded "parts," we can return to the separation of qualities, such as brilligance, brownness, or bravery. The mathematical properties defining a sharp partition, however, do not depend on what we are partitioning, except that they assume that we already know how to partition a set from the rest of the world. If we have succeeded in making *that* partition correctly, then three mathematical rules can tell us how to make further subdivisions.

A partition may be described by a set of ordered pairs. In the case of partitioning a set of states, for instance, each pair is two observed states. These pairs, in the case at hand, do not represent the relation "is a successor of," but the relation "has the same value of brilligance as." Thus, since (a, d, h, i, o) all have the same brilligance, we would find all possible pairs, the Cartesian product, of these states in the relation describing brilligance.

Clearly, a quality or property is not going to satisfy our idea of a quality if we cannot consistently identify it with a particular state. If, for example, we see state d and the inventor tells us it does not have the same brilligance as the last time—that the pair (d, d) is not in the set— then the whole idea of quality breaks down. This desire for consistency leads to the first mathematical condition for describing a partition, and thus, in our terms, describing a quality:

For every state, x, the pair (x, x) must be in the relation.

The mathematicians call this the *reflexive* condition. (Think of a *reflection* in a mirror.) When the partition describes a quality, it means that the quality cannot shift back and forth with time while the state remains the same. If we start with an idea of qualities, as we did with (*R, G, W*), we *identify* states by just such shifting of quality values. If we start with a holistic perception of states, we must *choose* our qualities so they have the reflexive property.

Checking on reflexivity may protect us against errors of absolutist thinking, of taking a relative property as an absolute one. For instance, if we attempt to divide all the people in a village into groups of "cousins," the attempt is faulty because the property of being a "cousin" is not an absolute property, but a relationship between two people. A person is not a cousin, but a cousin *of* somebody. The flaw in our thinking is exposed when we notice that obviously a person is not his own cousin. We therefore cannot partition the people of a village into groups that have various values of the quality "cousin."

The second property a relation must have to fit our intuitive notions of a quality is *symmetry*. The judgment as to whether two states are the same or different with respect to the quality in question should not depend, for example, on the order of appearance. If we first ask, "does *d* have the same brilligance as *h*?" and then ask, "does *h* have the same brilligance as *d*?" we expect to get the same answer to both questions. Psychologically, symmetry does not always hold, and judgments of sameness or difference may depend on the order in which the judgments are presented. In that case, the property in question does not uphold our usual idea of quality.

Again, symmetry may be violated in situations that seem at first quite innocent. If, for instance, we are trying to partition a village into groups of "friends," we may force the reflexive condition to be met by defining each person to be his own friend. But if we ask *A* if he is a friend of *B*, we may get a different response than if we ask *B* if he is a friend of *A*. By the Principle of Indifference, we have no a priori way of knowing which one to believe. Thus, if we try to build a theory on the property of being "in this clique of friends," we may become entangled.

Even if "friendship" is a symmetrical relationship in a particular system, we might still be unable to partition the system into subsystems of "friends" because of the necessity for *transitivity,* our third condition. Transitivity holds, for example, if we can say that *B* is a friend of *A*, and *C* is a friend of *B*, then *C* must be a friend of *A*. Clearly, in the case of friendship, transitivity need not hold at all, since *A* and *C* could just as well be either strangers or enemies.

Transitivity errors are the most frequent cause of improper discussions of qualities or parts. Consider, to take a common case, the idea of color classes, such as we call "red," "blue," or "green." How might an observer partition a set of color samples? He could examine one pair at a time to decide whether the two colors are the same or different. But when he has finished judging all the pairs, will the colors be partitioned? Perhaps—but it is not very likely. Why? Because the human color-sensing apparatus has some minimum color difference that it can detect, called the "just-noticeable difference," or JND by psychologists—or "grain" or "resolution level" by systems thinkers.

When there is a JND, *A* and *B* may be different colors (in the sense that a superobserver could distinguish them), but classed as "the same" by our observer. Suppose that *B* is just "a little more blue" than a green *A*, but not enough for our observer to notice, and that *C* is just "a little more blue" than *B*, but again, not more than the JND. It may happen that the difference in "blueness" between *A* and *C* *is* enough for our observer to notice, so that although *A* is the same color as *B* and *B* is the same color as *C*, *A* is not the same color as *C*.

Resolution levels are part of any measuring process, whether the instrument be machine or sense organ. With graininess, transitivity may not hold, and wherever transitivity does not hold, there is no complete partition, no clear division of a system into subsystems, no clear separation of system and environment. If we were measuring lengths of critical parts by comparing each one with the next, and the resolution level was 0.0002 inch, 1.0000 would be the "same length" as 1.0001, which in turn would be the "same length" as 1.0002, and so forth. But 1.0000 would *not* be the "same length" as 1.0002. Small wonder such measurements are made against a single standard, rather than successively against the last part measured.

Where the measurement is along a dimension not as simple as length, transitivity errors are even more frequent. In biology, for example, species have sometimes been defined using the concept of mating compatibility. If a male of one group could successfully mate with a female of the other and produce fertile offspring, the two groups were said to be of the same species. Though this seems a clear idea, on deeper examination it becomes as murky as earlier species concepts that depended on similarity of physical characteristics.

Along the Appalachian ridge each neighboring group of frogs can be successfully mated, but the frogs taken from the two ends of the chain cannot mate at all. Thus, to speak of dividing these frogs into species according to the mating criterion is to speak in fallacies. In fact, many naturalists have come to doubt the usefulness of any global species concept.

Although such a species concept is supposed to be an act of partition, we can view it as an error of *composition*. The definition of species that works so well on the local level fails when a large system, composed of many local levels, is studied.[10] The mating criterion fails to partition because it assumes a clarity of definition—"successful mating"—that does not correspond with the system modeled. In careless speech, it is easy to imagine that two groups either mate successfully or they do not. We do not often have our excessively sharp view of the world challenged by phenomena like Shasta the liger at the Salt Lake City Zoo, whose father was an African lion and whose mother was a Bengal tiger. For zoologists, who examine the world of animals more closely, the significant measure is of necessity more precise, taking into account that matings may be more or less successful, across an entire spectrum. Litter sizes may differ, number of survivors from the litter after some period may differ, number of litters per year may differ. Without considering a great variety of measures of reproductive success of a mating, it is impossible to derive a consistent and useful definition of species.

Well, not quite impossible. After all, if we take an arbitrary pair of animals—a frog and a horse, a dog and a wren, or an alligator and a duck—chances are they would not produce any offspring at all. We believe this even though, in all likelihood, nobody has ever tried to mate an alligator and a duck, at least without one eating the other. Yet we are probably right, even though the reason has nothing to do with the clarity of our mating concept.

As we shall see presently, the world often divides more clearly into parts than we would suspect from random divisions—divisions that have not been subjected to the tests of reflexivity, symmetry, and transitivity.

The Strong Connection Law

Clear logical thinking requires that we vary only one factor at a time.

A science textbook

All other things being equal, ...

Another science textbook

There is a world of difference between a mathematically correct partition and a scientifically useful one. For example, diseases are commonly described in terms of a heterogeneous classification, including

1. The invading agent (influenza virus, tapeworm)
2. The body's immediate response (rheumatic fever, cholera)
3. The ultimate damage (infantile paralysis, muscular dystrophy).

The working physician, of course, has to make do with what he can observe—he has no hope at all unless he can discriminate one disease from another. But the medical researcher seeks other ways of categorizing disease, ways that will not necessarily be so easy to recognize, but which can form the basis for going beyond the immediately given diseased states.

"Varying one factor at a time" is a useless admonition unless we have already partitioned into factors, or qualities. Yet we only discover what are useful factors by an enormous amount of experimentation with alternative ways of transforming one view into another. Once we have the "right" set of factors—once we make the "right" separation between system and environment—the solution we seek becomes simple. Indeed, this simplicity is what defines "right set of factors."

If we remain at the level of ostensive definition, we have *no* set of factors to vary—just this state, that state, the other state. Because certain sets of transformations used by sciences seem so "natural," some researchers think that separating into factors may be done in an arbitrary way. These are the people who "divide everything into two kinds of things."

A typical example (which shall remain mercifully anonymous) goes as follows:

1. There are two kinds of systems, large and small.
2. Within each category, there are two kinds of systems, centralized and decentralized.
3. These categories further break down into public and private.
4. There are the systems with mechanization and those without. . . .

We may be spared the remaining details: the pattern is clear. The technique is guaranteed to make a true partition, in the mathematical sense. But of what use is it?

Partitions, to be useful, must be dynamically useful. What difference does it make if the system is large or small? Centralized or decentralized? Public or private? It must be demonstrated how these categories partition the systems along lines that will help us in thinking about them. Only by trying out the process of varying one of these factors at a time do we discover whether they deserve the name "factors," or "properties." By the Principle of Invariance, it is the transformations we try, those that preserve and those that destroy, which teach us the meaning of a particular factor, or property.

Though we can never say *precisely* what we mean by a particular property, the more transformations we investigate, the more we feel we understand it. By "more" we mean something vague, but certainly not mere quantity. For instance, in Figure 5.8, one transformation would be to increase the dimensions of each box by 0.000001 millimeter, while another would be to increase it by 0.000002 millimeter. In this way, we could generate a ton of transformations without adding an ounce to understanding.

Nevertheless, though we cannot characterize exactly what we mean by "more" transformations, we do have some idea of what we might mean by "fewer." We may even consider the case in which there were *no conceivable transformations* that preserved a particular property. In that case, by the Invariance Principle, that property is synonymous with "existence," and in a certain sense is not a "property" at all.

Is this not the archetypical "system" property—one that is lost if the system is changed in *any* way? Is this not what we mean by "a property of the whole"? And do we not mean by "system" something that has *only* properties of this sort, something to which *no* transforformation can be applied without change?

At last we are near our goal. In the previous chapter, we demonstrated the consequences of carrying the strategy of reduction to its limit. Now we have made the same demonstration for "holistic" thinking. All that remains is to codify what we have discovered into a general systems law, *The Perfect Systems Law*:

True systems properties cannot be investigated.

In other words, the systems thinker, like the scientist, is in search of a Holy Grail—a perfect system—that could not be studied if found. Just like the scientist, or the poet, he is questing for an approximation to "truth," an approximation that he can never complete.

So let us retreat from this perfectionist position and see what we can salvage of what we have learned. The scientific revolution, based on a strategy of reduction, has made enormous contributions to our understanding of the universe. In doing so, it has worked well on certain systems, poorly on others, and remains untried on many more. To revive our handyman analogy, science has been working in the same neighborhood for a few hundred years. It has a particular box of tools, transformations to apply in the search for invariant properties, with which it has been able to solve many of the local repair problems. But after a while, we begin to have a *residue* of problems—the ones the handyman cannot fix with his particular box of tools. The systems researcher sees the *residue,* the situations in the world that science cannot, or has not, brought under its control.

This residue consists of two parts. First, there are those situations in which present scientific methods *could* work, but have not, either because they have never been tried or because they have been tried without proper imagination and understanding. Not everyone can repair a leaky faucet, even with the proper set of tools. Second, there are those situations in which the present tool kit will prove insufficient. These second situations are the proper concern of the general systems movement.

Is this a sterile dichotomy? Situations do not come labeled First Situation, Second Situation. Only by trying the tools in the scientific kit can we discover which are which. But since science has been working this neighborhood for a while, we have the right to assume that among the unsolved problems, the percentage of Second Situations has increased. To make another analogy, after we have been fishing in a small pond for a while, most of the easy fish will have been caught—and it may be time to change bait.

By an analogous argument, we may see that as time goes on, systems that are easily decomposable will have been decomposed, leaving systems that tend to be more difficult to decompose, systems that are more tightly connected. This, we believe, accounts for a portion of our impression that not any arbitrary collection—or "granfalloon," in Vonnegut's terms—should be called a "system." Nothing really prevents us from calling weakly connected collections systems, but such collections will be more easily broken into factors and thus yield their secrets to rather extreme methods of decomposition. Furthermore, most of them will have already been broken.

We can summarize these arguments in terms of the *Strong Connection Law*:

Systems, on the average, are more tightly connected than the average.

In other words, the elements of a system seem to be more closely integrated than elements of a "granfalloon." The people who come from Indiana do not impress most Americans in the same way that the people of the "International Communist Party" do. They are not, we believe, working in a tightly knit conspiracy to elevate the state of Indiana over the state of Illinois. Therefore, we have already mentally decomposed the Hoosier granfalloon into separate individuals—and they, not the grand set, form our systems.

The Strong Connection Law can be stated in several interesting ways. We can say, for example, that

A system is a collection of parts, no one **of which can be changed.**

Hazel the Hoosier can do many things that have not the slightest effect on the people back home in Indiana, but we imagine that when Ivan the Communist sneezes, the cell members in New York blow their noses.

We do not mean to state literally, by using this particular form, that a system is a Perfect System. We merely call attention to the property of interdependence that this collection is likely to exhibit, at least to casual or conventional observations. Therefore, such a system—as given—is not yet ready for the "one-factor-at-a-time" strategy, since the usual factors are likely to have been tried and found wanting. We have tried to understand the behavior of someone by applying our usual categories, but we have not succeeded. Therefore, we see a man as a "Communist," an inseparable part of a much bigger system, and not as the independent person he seems to be.

"Varying one factor at a time" is the same as "holding all factors but one constant." Therefore, the philosophy of decomposition can also be expressed by inserting the phrase "all other things being equal . . ." in the "if so" clause of our scientific laws. Corresponding to this guise of decomposition is another statement of the Strong Connection Law:

In systems, all other things are rarely equal.

Tracing arguments which led to the Strong Connection Law, we find that they originated in our need to simplify the world. If our mental powers were unlimited, we would not need to decompose systems into parts or properties, and the Strong Connection Law would not trouble us. Therefore, at least one source of the "system" feeling lies in our limited eye and brain capacity.

QUESTIONS FOR FURTHER RESEARCH

1. *Language Learning*

When a native speaker of English learns French, he usually experiences some difficulty in remembering which words are "masculine" (*le*) and which are "feminine" (*la*). Is it "La *plume de ma tante"* or "Le *plume de ma tante"*? *Tante* is clearly feminine, because we know that all aunts are women, but what about *plume*? Well, perhaps a featherpen "seems" feminine. Then what about an automobile? The French Academy is reputed to have debated for 40 years over whether it was "le *voiture"* or "la *voiture."* How does an English speaker learn the sex of a feather? How does a French child learn the same thing? How does the French Academy know the sex of automobiles?

2. *Developmental Psychology*

From where do we get our concept of a "thing" or "object"? Are we born with it, or does it develop? Research on this question could begin with the following article:

T. G. R. Bower, "The Object in the World of the Infant." *Scientific American,* **255,** No. 4, 30 (1971).

3. *Unidentified Flying Objects*

From time to time, particularly in the heat of the summer, we get a flurry of reports that mysterious flying objects have been seen glowing in the dark, flying at fantastic speeds, executing unbelievable maneuvers, and so on. We are not concerned with such objects here, but with another question. What is the possibility that we are being visited by extraterrestrial "beings" that are *not* strongly connected in a way that we would recognize them as "objects"? In other words, are there "Unnoticed Flying Systems"?

4. *Political Anthropology*

According to Fortes and Evans-Pritchard, there are two major classes of political systems—states and stateless societies. Among the stateless societies, there are two kinds (not named): one categorized by corporate lineage segments, the other distinguished by a nonsegmented kinship structure.

How would an anthropologist decide whether this was justified or unjustified (fertile or sterile) dichotomizing?

Reference: M. Fortes, and E. E. Evans-Pritchard, *African Political Systems,* pp. 6–7. London: Oxford University Press, 1940.

5. *Psychology and Philosophy*

Another area in which JND's are found is in the perception of time, that is, in the perception that two events take place "at the same time." (See, for example, George A. Miller, *Language and Communication,* revised ed., pp. 47–49. New York: McGraw-Hill, 1963.) The "psychological present" corresponds to the span of time—the JND—in which two events can occur and not be recognized as being "at different times." What are some of the possible ramifications of this JND on psychological theory? What about on physical theory, which is based in some part on our conception of simultaneity?

6. *Linguistics*

Language families are often described as if it were possible to separate, say, French from Italian. Are there sharp boundaries between such languages? If not, what is found at the interface—in the Alps, for instance, where two language groups interact? What implications do these observations have for language theory? For language training?

7. *Intransitive Gambling*

There is a children's game called "scissors–paper–rock" in which the two competitors simultaneously thrust out one hand with the sign of scissors (two fingers), paper (flat hand), or rock (closed fist). The winner is decided by the intransitive rule: "scissors cut paper; paper covers rock: rock breaks scissors." In such an intransitive game, there is no "best" play, which is a difficult concept for children to learn.

But it is also difficult for adults to learn, as Martin Gardner demonstrated in his Mathematical Games column in which he presented the dice game designed by Bradley Efron, of Stanford. There are four dice, marked on their faces with the following sets of numbers:

$$A: (0, 0, 4, 4, 4, 4)$$
$$B: (3, 3, 3, 3, 3, 3)$$
$$C: (2, 2, 2, 2, 6, 6)$$
$$D: (1, 1, 1, 5, 5, 5)$$

The first player selects any die from the set, then the second player may choose his die, after which the selected dice are thrown and the winner is the one showing the higher number.

Make up a set of such dice by marking numbers on the faces of ordinary dice, and play the game until you understand why even though the first player gets to choose the "best" die, the second player has two-to-one odds in his favor! Clue: Gardner points out that "the paradox (insofar as it violates common sense) arises from the mistaken assumption that the relation, 'more likely to win,' must be transitive between pairs of dice."

Reference: Martin Gardner, "Mathematical Games." *Scientific American*, December 1970, Vol. 223, #6.

8. *Survey Research*

When surveys are made, the individual questionnaires are usually subjected to various tests to determine their "validity." Among the tests that are applied is the test for transitivity among certain stated preferences. For example, in a study of optimal family size, women were asked such questions as:

"Which would you prefer, to have no children or an only child?"

"An only child or two children?"

"Seven children or eight children?"

The "preferred family size" is then derived from the responses. If, however, a woman says something like, "I prefer one to none, two to one, three to two, and four to three, but I would rather have no children than four children," her questionnaire is discarded as "not valid." Discuss the assumptions underlying this procedure. Give examples of when it might be justified and when it might not.

9. *Civil Engineering*

Canals and waterways are often thought of as penetrating barriers or boundaries, but the penetration is differential. For example, the Suez canal, though a sea-level canal without locks (which people recognize as "barriers"), has the Bitter Lakes with their high salinity to stop many potentially crossing species that cannot survive long enough in such an environment. To us, it just looks like "water" in the canal, and "fish" live in water. Some "fish," of course, can move through this "barrier" very easily, while others not at all. In the middle are many who have varying probabilities of moving through.

Discuss the ways in which different concepts of a canal as "boundary" or "interface" will affect the systems on the two sides. For a specific example, consider the proposals for a new Atlantic–Pacific canal to supplement the Panama Canal, but without locks.

> Reference: William I. Aron, and Stanford H. Smith, "Ship Canals and Aquatic Ecosystems." *Science,* **174,** No. 4004, 13 (1971).

10. *National Parks*

The national parks in the United States were set aside in an effort to preserve representative parts of the country in their "pristine" condition. The boundaries of the parks, however, were often drawn along lines of scenic beauty rather than ecological coherence, so that many of the parks do not represent "complete" ecosystems, in the sense of maintaining a cycling of materials and energy in a closed manner, or rather, in the manner existing before modern man came along. For example, a migratory mammal or bird may spend only part of its yearly cycle within the boundary of a park, or a river may not have its source within the park boundaries. Discuss other examples where the artificial boundaries set by man's naive conception of "beauty" do not correspond with ecological realities, as well as circumstances where a correspondence might be expected. Discuss the consequences of some of these differing boundaries, and ways in which the artificial boundary becomes "real."

Reference: F. F. Darling and N. D. Eichhorn, *Man and Nature in the National Parks*. Washington, D.C.: Conservation Foundation, 1967.

11. *Markets*

The *market* is a boundary, or part of a boundary. The market is the condensation of all transactions across the boundary into a single visible arena. This concentration permits the regulation of the transactions that might otherwise be fraught with danger for both systems, since, for example, there may be cultural clashes. Markets are found to be laden with rules circumscribing behaviors. Discuss the functional significance of some of these behaviors for the maintenance of the interface.

12. *Ports*

The market, as described above, is an example of the more general phenomenon of a "port," that is, a special place on the boundary through which input and output flow. Over *most* of the boundary, no exchange, or very limited and perhaps unavoidable exchange can take place. Only within the location of the port can the dangerous processes of input and output take place, and by so localizing these processes, special mechanisms may be brought to bear on the special problems of input and output. Compare various examples of ports, such as the mouth, a port city, a computer terminal, or a door.

13. *Membranes*

In contrast to the concept of port, or localized interface, is the concept of "membrane," or distributed interface. An obvious example is the cell wall that may be penetrated at almost every point on its surface, but not by everything and not at all times.

Compare various examples of membranelike interfaces, such as cell walls, the border between Nebraska and South Dakota, the skin, or the walls of a tent. Contrast the concept of port with the concept of membrane.

Reference: Lawrence I. Rothfield, Ed., *Structure and Function of Biological Membranes*. Chicago: Academic Press, 1971.

14. *Local Taxonomy*

Each different locality on earth has its own unique set of flora and fauna, and thus might be expected to have its own local method of taxonomy. Linnaeus, who gave us the beginnings of the system now used by biologists, was essentially codifying the folk taxonomy of a particular region of Europe in 1737. Linnaeus worked to obtain genera that were separate, distinct units with natural limits on which all could agree. But he was working with only a few thousand of perhaps ten million species of organisms in the world, of which a mere 10% or 15% are described at any level even today. Discuss the problems that may arise in scaling up the Linnaean system to a global system of taxonomy.

> Reference: Peter H. Raven, Brent Berlin, and Dennis E. Breedlove. "The Origins of Taxonomy." *Science*, **174** (17 December 1971).

15. *Pharmacology*

In our thinking about medicines and other drugs, we often speak of "the effect" of a drug as if it were some fixed entity separable from all other such entities. As long as taking drugs is an infrequent event, this view may be adequate enough, for few occasions will arise on which two drugs are being taken at the same time. As drug use increases—as we move closer to a "one-problem, one-pill" philosophy of life—such a clear separation may no longer be a viable model of drug effects. For example, alcohol often has a strong interaction with barbiturates and tranquilizers, an effect well known by now because we have so many alcoholics in the general population. But more recently, interactions among prescription drugs has become a matter of serious medical concern. Research such interactions, and discuss measures that could be taken to prevent dangerous consequences. Relate these measures to the Square Law of Computation and the Strong Connection Law.

16. *Welfare Economics*

So-called "welfare economics" is an attempt to answer the question of what economists mean by one state of the economy being "better" than another.

A Pareto optimum is attained when the system is no longer in a state where it can move to a new state in which no person is worse off and at least one person is better off (always in their own estimation).

Contrast this with Veblen's idea that there is no change that does not help someone and hurt someone else at the same time. If Veblen is right (and his is a "system" idea), then Pareto's optimum is a meaningless construct. Still, might it be meaningful in an approximate sense? Discuss these two concepts in terms of the Strong Connection Law and the Square Law of Computation.

> Reference: Vilfredo Pareto, *Manuel d'économie politique*, 2nd ed. Paris, 1927.

READINGS

RECOMMENDED

1. T. G. R. Bower, "The Object in the World of the Infant." *Scientific American,* **225,** No. 4, 30 (1971).
2. Peter H. Raven, Brent Berlin, and Dennis E. Breedlove. "The Origins of Taxonomy." *Science,* **174** (17 December 1971).

SUGGESTED

1. Oskar Morgenstern, *On the Accuracy of Economic Observations.* New Jersey: Princeton University Press, 1963.
2. Hermann Hesse, *Magister Ludi,* In *Eight Great Novels of H. Hesse.* New York: Bantam Press, 1972.

NOTATIONAL EXERCISES

1. Given the set (A, B, C, D), do the following pairs form a partition?
$$(A, A) \quad (A, B) \quad (B, A)$$

2. Do the following pairs form a partition?
$$(A, A) \quad (B, B) \quad (C, C) \quad (D, D)$$

3. The following?
$$(A, A) \quad (B, B) \quad (C, C) \quad (D, D) \quad (A, C)$$

4. What pair would have to be added to the set of pairs in Exercise 3 to make it a partition? What would be the three "parts"?

5. Suppose the pair (A, B) were added to the partition of Exercise 4. What additional pairs would have to be added to make a partition? What would be the parts?

ANSWERS TO NOTATIONAL EXERCISES

1. The given set of pairs does not form a partition because it lacks the reflexive property since, for example, (B, B) is missing.

2. Yes, because the reflexive property is satisfied and the other two properties are satisfied by default. Remember, for example, that the symmetric property says that *if* (A, B) is in the set, (B, A) must be in the set. Since there are no such heteroliteral pairs, the condition is said to be satisfied by default, and the partition is into the set of "individual" parts.

3. This is not a partition because although (A, C) is present, (C, A) is not, so symmetry is not satisfied.

4. By adding only (C, A), all three conditions become satisfied, and the partition is effectively

$$\{(C, A), B, D\}$$

5. If (A, B) were added, then (B, A) would have to be added by symmetry. But then, since (B, A) and (A, C) are in the set, (B, C) must be added to satisfy transitivity, after which we must obviously add (C, B). This gives the partition

$$\{(A, B, C), D\}$$

6

Describing Behavior

In my own case, pursuit of operational analysis has resulted in the conviction, a conviction which has increased with the practice, that it is better to analyze in terms of doings or happenings than in terms of objects or static abstractions.

P. W. Bridgman[1]

Simulation—The White Box

No artifact devised by man is so convenient for this kind of functional description as the digital computer. It is truly Protean, for almost the only ones of its properties that are detectable in its behavior (when it is operating properly!) are the organizational properties. The speed with which it performs its basic operations may allow us to infer a little about its physical components and their natural laws; speed data, for example, would allow us to rule out certain kinds of "slow" components. For the rest, almost no interesting statement that one can make about an operating computer bears any particular relation to the specific nature of the hardware. A computer is an organization of elementary functional components in which, to a high approximation, only the function performed by those components is relevant to the behavior of the whole system.

Herbert A. Simon[2]

In previous chapters we have discussed the "black box"—the system that could only be known through observing its behavior. To some minds, all systems thinking starts with the black-box paradigm, but to others, an exactly opposite approach is taken. For us, neither approach need be worshipped as a religious icon. The black box is one approach to understanding; the "white box," or simulation, is another. We must understand each if we are to understand the other.

Some systems thinkers view simulation as the ultimate tool because they believe that the way to demonstrate understanding of behavior is to *construct a system* that exhibits that behavior. Instead of the inside of the system being *perfectly hidden,* it is *perfectly revealed*—a white

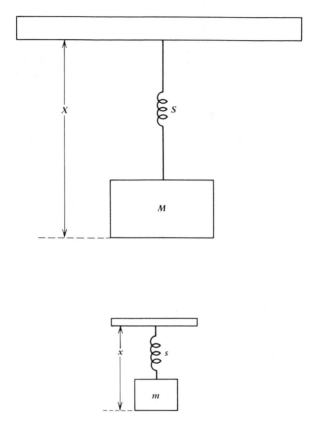

Figure 6.1. A physical system and a scale model.

box rather than a black box. As we shall see, however, because of our own limitations, no box is ever entirely revealed to us, even if we construct it ourselves. Even the simplest systems sometimes contain surprises for their builders.

Nevertheless, if we can assemble a system that appears to behave in the same way as a system we claim to understand, our claim, in turn, gains in strength. At least if we try to assemble such a white box and fail, our claim is thereby weakened. But, we can never be sure that our simulated system captures all the properties of the studied system. To do that, we should have to perform an infinite number of transformations.

Systems may be simulated by building scale physical models, as is done in designing ships or planes. Such simulation has immediate intuitive appeal, for a model ship "looks like" a ship. Nevertheless, model builders must work hard to overcome their intuition. In the early 1900s, for example, when geologists started making small-scale physical models of rock systems, they naturally chose to represent shale by shale. But because of the scaling involved, the properties of the model shale were not appropriate, being too hard, for one thing. Eventually they learned that the best model analog for shale was asphalt, a better scale shale than shale "itself."

Only explicit scaling laws enable modelers to predict the relationship of the behavior of the model to the behavior of the ultimate ship, plane, or mountain. The study of such scaling laws is known as "dimensional analysis," and is particularly recommended to those would-be systems thinkers with appropriate mathematical training.[3] For instance, in Figure 6.1, we see diagrams of two mechanical systems, one large and one small. Each system is a mass *m* hanging from a spring. Such a system will tend to oscillate at a regular frequency, and could be used, for example, as the timing source in a clock.

Now suppose we were planning to build a clock on this principle for the top of a tall steeple. If the clock were to be big, expensive, and hard to adjust once it were in place, we might want to use the small system to simulate the large one before actually building it, so we would know, for example, how big a mass to use with a particular spring. We could apply dimensional analysis to translate observations on the small model into predictions about how the actual clock would behave. The clock will be like the model, only "bigger."

A somewhat less intuitive form of simulation is *analog computing.* Most analog computing today is done with *electrical* analogs, electrical circuits that are in some way similar to the system under study. Figure 6.2 shows the electrical circuit diagram of an analog of the mechanical

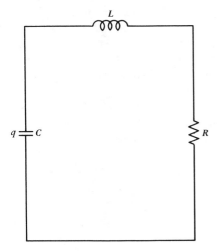

Figure 6.2. An electrical analog to Figure 6.1.

system of Figure 6.1. To the inexperienced eye, the similarity between these systems is not manifest the way it is between a mechanical system and its scale model. Nevertheless, the engineer can see a strict analogy, or isomorphism, between the two, just as readily as we saw the analogy between the small spring and the large spring. In fact, the analogy with the electrical circuit is ordinarily easier than with the scale model. The transformation laws are more direct, requiring no sophisticated dimensional analysis.

For instance, the electrical charge q on the capacitor C is the direct analog of the distance X in the mechanical system; the electrical capacitance of the capacitor corresponds directly to the elasticity of the spring; the resistance of the resistor R corresponds to the friction in the mechanical system; and the inductance L corresponds to the mass m. Therefore, by observing the charge on the capacitor, we can make conclusions about the movement of the mass, without building any clock at all.

Much more complex systems—biological, mechanical, or what have you—can be simulated by electrical circuits in a like manner. But since most of us do not meet the requirements in terms of electrical engineering skill, analog computers are not a serviceable kind of simulation for fledgling systems thinkers—though, again, those with adequate preparation will reap rewards from their study.[4]

Fortunately, the *digital* computer provides a more accessible simulation tool to those without training in physics or electrical engineering. The digital computer has a number of other practical advantages as a general simulation tool over scale models and analog computers, but only one need concern us here. The relevant feature is "programming," the scheme that permits us to build white boxes in a language more or less natural to all of us, so that we may all be on equal footing in our discussions. The study of computer programming is an excellent way to improve the quality of one's systems thinking[5], but programming experience will not be needed to understand a procedure rendered in a language resembling real computer programming languages.

For example, suppose we want to simulate those music boxes you saw in the warehouse, so they could be studied when the inventor was not around. Our computer can be hooked up to a pair of lights and a whistle, or better yet, to a typewriter that could type values of (R, G, W). We can enter into the computer's memory the 20 states to be displayed, which we could designate as $(S_1, S_2, \ldots, S_{20})$. Then we could furnish the computer with a *program*—a set of instructions telling how to control the typewriter. The program might look something like this:

1. Repeat line 2 20 times, moving i from 1 to 20.
2. Display state S_i.

This program would cause one complete cycle of states to be typed. If we wanted to repeat the cycles indefinitely, we could insert the following statement:

0. Repeat lines 1 and 2 indefinitely.

By entering this three line program into our computer, we make it slavishly continue to type the 20-state cycle until we tire of looking at it or run out of paper.

The perceptive reader will now point out that the above model is a form of cheating, for to make it, we have to understand very little about music boxes. The whole technique resembles a chess-playing machine with a midget grand master hidden inside. In place of the midget, we have hidden the 20 states in just the right order. Small wonder, then, that they come out exactly right. What does all that prove about our understanding of the music box? The answer, of course, is "nothing."

The only way we can acquire confidence in our understanding is by putting in *less than* the 20 states. Then we wait to see if the resulting behavior matches our observations of the real box. For instance, in Humpty Dumpty's box, we were able to discriminate two independent cycles, one of lights with four states (L_1, L_2, L_3, L_4) and one of tones with five states ($W_1, \ldots, W_5$). If we store these nine substates in the computer's memory, we can employ the following program:

1. Repeat line 2 indefinitely, changing i from 1 to 4 and j from 1 to 5.
2. Display state (L_i, W_j).

This program is not quite as bogus as the previous one, since we have put in only 9 states and gotten out 20. Why? Because we had been able to decompose Humpty Dumpty's box into properties that, unlike Humpty Dumpty, *could* be put back together again.

It is in this sense, then, that we say that simulating a system demonstrates understanding. If we do not simply *copy,* or *mimic,* the system, but assemble a model from smaller numbers of parts, or states, or properties, then we must know *something* about the system.

To demonstrate this constructive concept of simulation, let us assemble a much more complex machine, complex in the sense of having numerous states. Let us suppose that each state is represented by 100 digits, giving a Cartesian product of 10^{100} members. Storing all the states of this machine would require a computer with a memory large enough for 100×10^{100} digits. Since no computer could have so much memory (there are, after all, less than 10^{100} particles in the known universe!), we are automatically certified not to be cheating, at least not in the sense of storing all possible states in advance. If our machine is to be simulated, we shall be forced to generate its behavior from a much smaller number of states.

Before we specify the simulation in detail, let us give it a plausible setting—an *interpretation* of the white-box model. Suppose we have just received a Ph.D. in General Systems Thinking and have opened an office downtown with our "General Systems Thinker" shingle over the door. Because there is such a shortage of general systems thinkers, we find a customer waiting at the door on our very first day, a gnarled little man with a bald, sunburned head and a white goatee. By way of introduction, he hands us his card, which says:

E. S. O'Teric, Grand Mystagogue
OCCULT
Organization of Cabalist Clubs
Using the Latest Technology

and explains his problem.

"You are familiar with our organization?"

"Well . . . ."

"No matter. We Cabalists have only recently opened our membership and begun to solicit recruits. For centuries we have been a tightly closed group, working out the numerical secrets of the universe for the benefit of all mankind. Recently, however, so-called "science" has begun to exert such an attraction on young minds with cabalistic potential that we have had to modernize our ways. Now we use the latest technology—computers, for example—and the latest organizational methods to form our clubs."

"Oh, yes. You had an advertisement in the Bay Guardian."

"Good. You read the Bay Guardian. You must be a modern systems thinker. Just what we need to solve our problem."

"But what is your problem?"

"You see, something is going wrong with our clubs, almost as fast as they are formed. We do not understand how, for the clubs are organized strictly according to the laws of arithmetic. For example, each club must contain exactly 100 members—a perfect square. Through the interaction among members, each club is supposed to strive to reach the state of perfect unity, but instead they become nothing, perfect ciphers."

"I'm afraid I don't understand."

"I will draw you a model. You see, as each club member joins, he is assigned to a caste, a Roman numeral, I through X, depending upon a combination of his Zodiacal sign, his IQ score, and his shoe size. Some are already I's—having achieved perfection naturally—and these we expect to be our leading caste, helping the others to reach their state of I-ness. Instead, their influence seems to be weak, and gradually the club members come to follow the know nothings—the "zeros," or X's. We have even gone to the extreme of excluding X's from membership, but their vacuous philosophy seems to arise spontaneously among the minds of modern youth. If we cannot stop them, the ancient lore of the Cabal will be driven from existence . . ."

"Why don't you sit down and have a cup of tea, so that you may explain things a little more calmly. I'm not sure, however, that general systems thinking can help you."

"It must help us. I've been told that generalists can solve *any* problem."

"We can *talk about* any problem. Solving is another matter. But don't worry. If you're not completely satisfied, your money is refunded in full. We may be a new profession, but we have our code of ethics. Now, what about your clubs."

"Well, we don't understand how the zeros—or X's, as we call them— have such an influence. The clubs never meet as a whole body, but only as individual pairs of members for face to face Cabalistic discussions. That way we hope to prevent the spread of perfidious nihilism by mere oratory. But pure reason doesn't seem to stop them."

"What do you mean by 'pure reason'?"

"Everything in the universe, as you, an educated person, undoubtedly know, is governed by numbers and the laws of arithmetic. Thus, in our clubs, members are not known by name, but by an Arabic number which they are assigned by National Headquarters. Each club therefore has members 01 through 100, and each member has his caste, I through IX, and I'm sorry to say, zero, which we call X because the Romans didn't have a zero. Each week Headquarters sends out a list of numbers of members who have been selected for meetings according to principles that I am not at liberty to reveal."

"I can't help you if you don't tell me everything."

"At the appropriate time, I will explain to you how the members are selected for meetings. Suffice it to say for the moment that, for all intents and purposes, each selection is of a random pair of member numbers. You know, of course, that nothing is really random?"

"Of course."

"Good. But to the uninitiated, these membership numbers *look* random and are given in the form of ordered pairs, such as (03, 17), (95, 08), (66, 45). The first member of the pair is the teacher, the mystagogue, and the other is the pupil, the catechumen. At the end of the session, through application of the principles of the cabal, the disciple has been brought to a new caste, and receives a new Roman numeral."

"You mean he gets the teacher's—the, uh, mystagogue's—caste?"

"No, not necessarily. You see, it's all determined by the rules of multiplication. For example, if the master's caste is IX and the novice's caste is VII, then the new caste is III."

"I'm afraid I don't follow."

"Arithmetic! Simple arithmetic! *All* the universe is simple arithmetic!"

"Yes, yes. Of course. But could you just explain it once more to be sure I understand?"

"Oh. I'm terribly sorry. It's the youth of today that have me in such a dither. Let me give you the entire rule. To find the new caste of the pupil, you multiply the castes of pupil and teacher together. If the number has two digits, you drop the first one. So, $9 \times 7 = 63$ and dropping the 6 gives 3. In other words, IX and VII produce III."

"I see. So if the teacher were VIII and the pupil were IV the new caste would be $8 \times 4 = 32$ and dropping the 3 gives II. Right?"

"Correct."

"And what do you want *me* to do?"

"I want you to simulate our clubs on one of those computers of yours, and to make us a report on what you discover. We cannot let the soulless armies of the night defeat the indivisible oneness of light."

And so, after agreeing on a fee and escorting Mr. O'Teric to the elevator, we begin to work on our simulation. We recognize that the rule for generating new castes is precisely one given by Ashby as a homework problem in an *Introduction to Cybernetics*[6], so we dig up our course notes and use them for the first time in our lives. We realize that in the computer's memory we can simulate each club member and his caste by a single "memory cell," which can hold one Roman numeral. For the whole club, therefore, there will be 100 cells for the castes of the 100 members, or $d = (d_1, \ldots, d_{100})$, where d is the caste number. We will also need a source of Arabic number pairs, (i, j), to select teacher (i) and pupil (j). These numbers might be stored on magnetic tape or on punch cards, given by us through the typewriter, or transmitted over the radio from some distant point via satellite. Since Mr. O'Teric did not want to reveal the formula for generating these numbers, we will simply have our program obtain them one pair at a time by such an instruction as

1. Get next pair (i, j).

We can try various schemes of generating (i, j), and later we can ask the Grand Mystagogue to supply us with numbers to check out our simulation.

Once the number pair has been brought in, it selects one pair of club members, member i and member j. Their castes, d_i and d_j, will then be multiplied to produce the product, which we will store in a temporary cell called t. We can instruct our computer to do this by writing

2. $t = d_i \times d_j$

Next, we want to change the old caste into the new caste by taking the last digit of t and storing it in place of the old value of d_j. This step can be written in the program as

3. $d_j =$ last digit of t

And that is all there is to it! The entire program may be written something like this:

O. Repeat lines 1–4 indefinitely.
1. Get next pair (i, j).
2. $t = d_i \times d_j$
3. $d_j =$ last digit of t
4. Display $(d_1, \ldots, d_{100})$.

Figure 6.3 traces the behavior of our program through two repetitions of lines 1–4. The first (i, j) pair happens to be (28, 35), which causes members with castes 3 and 7 (III and VII) to be selected. Next, t is given the value $3 \times 7 = 21$. The last digit of t, 1, is then removed for storing in place of d_{35}, since $j = 35$. This causes the caste in d_{35} to be changed from VII to I, which, of course, changes the state of the entire system.

In a similar manner, the next input pair (38, 30) causes d_{30} to be changed to VIII, as you should verify for yourself. In this manner, changing one caste at a time, the system may travel indefinitely far from its initial state. The behavior of the system through 20 transitions is seen in Figure 6.4, which you should follow until you have mastered the scheme, or "algorithm," for the transition from state to state.

Actually, for ease of presentation, Figure 6.4 shows not our original system of 10^{100} states, but a simpler system of 10^{40} states. We can easily understand this system, and easily translate what we learn to the 10^{100} system, because we know the internal structure of the computer model.

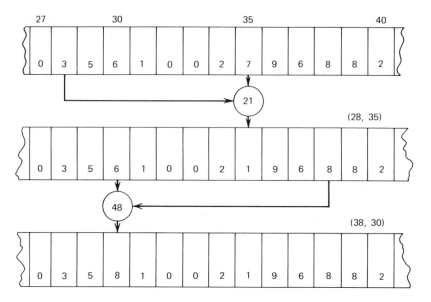

Figure 6.3.   The structure of the OCCULT simulation.

The two models are not *exactly* the same, but they are similar enough
that we know precisely how to change the program for one into the
program for the other. In effect, all we have to do is change line 4 to
read:

4. Display $(d_1, \ldots, d_{40})$.

Better still, we can make line 4 read:

4. Display $(d_1, \ldots, d_n)$.

and start the program by furnishing a value for n that will determine
how big a club we want to simulate. Such a variation of n is precisely
the kind of thing an electrical engineer does in experimenting with different values of the resistance R or the capacitance C in the analog
computer of Figure 6.2. We call this operation "varying a parameter,"
and it permits us to convert the simulation of one system to the
simulation of a *similar* system. The way in which the models are
similar depends upon how the parameter is involved in the program.

In the analog computer of Figure 6.2, changing the resistance
parameter would not change the *structure* of the model. In analog computers, changing structure is more difficult than simply changing the
value of resistance—the wiring among the components must be
changed. To change structure in the digital computer, the program has
to be changed. But, since the program is stored in the computer's

memory, the change in structure can be accomplished more readily, as long as the *kind* of structure change has been anticipated.

In our OCCULT System, the number of club members can be changed simply by changing the value of *n*. One possible hypothesis is that the number 100 has something to do with the spread of nihilism among the members of a club, so, by making the number of members a parameter, we permit ourselves to change the club size and see if that

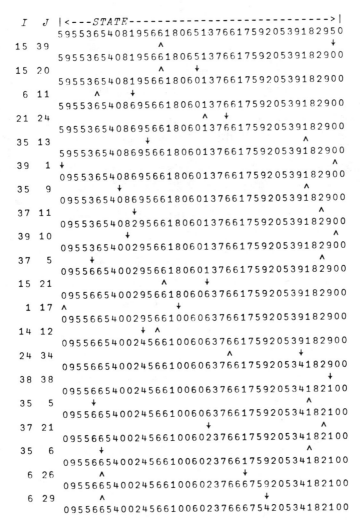

Figure 6.4. Sample output from a simulation for a 40-member OCCULT club.

affects the behavior of the club. Perhaps we could earn our fee simply by advising Mr. O'Teric to change his perfect square to some other number, in order to conquer the forces of darkness. Unfortunately, as we shall discover, it is not quite that easy.

State Spaces

A place for everything, and everything in its place.

A well-known aphorism

As soon as we begin working with a system of 10^{100} states, we feel the need for new metaphors, or reconditioned versions of old ones. With a 20- or 24-state music box, we could readily represent behavior by writing down letters and connecting them with arrows to indicate movement from state to state. How the states were placed on the page did not matter much. But with 10^{100} states, we are going to have to be a bit more orderly. In the first place, we shall be unable to write small enough to get 10^{100} states on a single piece of paper, and even if we could, we could not see anything we might be looking for. We need a *systematic* arrangement, and one that takes up very little space per state.

If the state of the system happens to be composed of *two* properties, such as (lights, whistle) or (brilligance, mimsy), we could arrange them in a tableau such as we see in Figure 6.5. Each square in the tableau

Mimsy

		V	W	X	Y	Z
	A	h	a	o	i	d
	B	v	g	m	n	k
Brilligance	C	j	b	c	p	s
	D	t	e	q	u	w

Figure 6.5.   States arranged in a two-dimensional tableau.

would then represent one and only one state, and lines could go from square to square to indicate change of state. If there are many values to each property, we can shrink the squares to points, in order to get as many as possible on the page.

Such a tableau is closely related to the Cartesian product, or product set, or "product space"—as it may be called to make the analogy with physical space explicit. It was Descartes who gave us this explicit formulation, and we find his name both in the set form—Cartesian product—and in the spatial form—Cartesian coordinates. This method corresponds to Descartes' dictum from the second maxim of *Discours de la Méthode*: "to break down every problem into as many separate simple elements as might be possible." We saw how the "properties" were broken out of the system by partition; the product space shows how they are put back together in a systematic way.

If each decomposition is truly a partition, then the product space must encompass the original possibilities entirely. In that case, there is a place for every state and every state has a place, the place being specified by one element of the Cartesian product, such as (B, X), in which we find state m, or (D, Z), in which we find state w.

From Descartes, we learned how to designate each point in a plane (a two-dimensional surface) with a unique *pair* of numbers. In fact, we call it "two dimensional" *because* it can be described by pairs. Of course, there is an infinitude of different ways of numbering the points with pairs. We may find a given corner in New York by the pair of cross streets (44th and First Avenue), by the number and street of a corner building (787 First Avenue), or by the distance and direction from some fixed point (2 miles Northeast of the Empire State Building). Any such system will do, but some are more convenient for specific purposes. If we forget that our grids in physical space are arbitrary, we will be surprised, on crossing the Equator, not to find a big red stripe.

We may find it convenient to establish a correspondence between the *states* of a two-variable system and the *points* on a physical plane, but we must not forget that this assignment, too, is arbitrary. For example, if the properties are finely grained, certain arrangements may appear to yield an unbroken or "continuous" line of behavior wandering about the plane. Yet this appearance of continuity may be entirely a consequence of our way of assigning numbers to the property values. Of course, the Principle of Difference may apply. We may *want* to find such an assignment, since it will reduce our mental effort at remembering and describing the behavior. Indeed, much of the work of the systems thinker is in assigning numbers to properties so that system behaviors trace out nice, neat figures in the planar representation.

When we do succeed in creating a point of view from which system behavior appears continuous, the arrows pointing from state to state may be thought of as becoming very, very tiny. In that case, we may have a concept of "nearness" of two states, so that *areas* on the plane may be taken to represent *sets of states,* or regions, that are somehow related to one another. *Topology*[7] is the branch of mathematics that studies how points of view can be transformed and still preserve such properties as nearness—but no amount of mathematical sophistication can mask the fact that the original nearness may simply have been put there by the observer.

We encounter two-dimensional representations of system states every day, and they are so familiar to us we may not notice them as such. (Figures 1.8, 1.9, and 2.2 were such representations.) Sometimes, they take a more static form, with areas not representing sets of states of one system, but a collection of single states of similar systems. In other words, instead of representing *one system at different times*—the so-called "diachronic view"—the points in the plane may represent *different systems at one time*—the "synchronic view." The method works

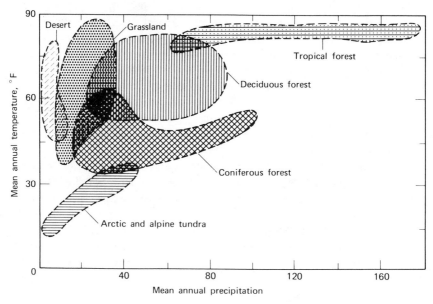

Figure 6.6. Six biomes characterized by two variables in a two-dimensional state space. (National Science Foundation)

equally well for either view, corresponding to the usual scientific method of substituting continuing observations on one system for many single observations on similar systems, and vice versa.

For example, Figure 6.6 is adapted from a National Science Foundation report on an ecosystem analysis program. The chart characterizes all the regions of the world as two-variable systems, the variables being

(mean annual temperature, mean annual precipitation)

Any existing region can thus be plotted as one state, or point, on this chart, since every location has some value of each of these variables. The shaded regions represent observed habitats or "biomes," three types of forest, plus grassland, tundra, and desert. We can see instantly from the chart that deserts are hot and dry—which we presumably knew already—and that tundra is cold and *dry*—which we might not have known.

Much of the value of this representation comes not from what is on it, but what is *not* on it. While everything may have its place, there may be places that have nothing in them, that is, combinations of properties that have not yet been observed. Such holes in the state space suggest to us that:

1. Our observations are not complete, and there are other states not yet observed.

or

2. Our categorization into properties is too broad.

The classical example of the first type is Mendeleyev and his periodic table of the elements: the holes led to the discovery of previously unknown elements. Figure 6.6 is probably an example of the second type, for we see that the lower right corner is entirely empty. Since the chart supposedly represents all actual places on earth, we see that there are no places that are both cold and wet—in spite of our impressions of Binghamton, New York. This empty category suggests that we look for reasons why such a combination is never found, or perhaps that we look for a different decomposition into properties, one that will fill the diagram more uniformly.

Simply laying out the states of a system in this systematic way may yield information we had not noticed in other representations. While it is easiest to use two-dimensional charts, we can also make three dimensional models of three-property systems, in which case, sets of states may be represented by volumes, not areas, though behaviors will still be represented by lines.

Our mental structure does not equip us to go beyond three dimensions in actual visualization—we cannot literally see a picture of a "four-dimensional cube" or other such artifact. When nonmathematicians hear talk of n-dimensional space, they are intimidated into thinking that mathematicians have some super mental powers, when in fact the only special power these mathematicians have is the power of *extrapolation*. They do not "see" n-dimensional space in any literal sense, but simply continue to apply the same mathematical operations without regard to the number of dimensions involved. A point on a two-dimensional plane is specified by two numbers; a point in a three-dimensional space is specified by three numbers; therefore, by extrapolation, a "point" in seven-dimensional space is specified by seven numbers. A one-dimensional object, a line, separates a two-dimensional object, a plane, into two parts; a two-dimensional object, a plane, separates a three-dimensional object, a solid body, into two parts; therefore, by extrapolation, a six-dimensional object separates a seven-dimensional object into two parts. Thus, we don't have to "picture" seven-dimensional objects to speak about them or to manipulate them.

An n-dimensional space in which each "point" represents a state of some system is called a "state space." We make imaginary operations on a state space, corresponding to certain operations in two- or three-dimensional physical space. For example, sectioning a three-dimensional body as for microscopy produces a two-dimensional slide. The analogous act in n dimensions has been called by various names—"sectioning," "projection," "dimensional reduction"—each corresponding to some particular physical analog. The space produced by this act is similarly called a "section," "projection," or "subspace."

Dimensional reduction may be forced on us by our instruments (the light microscope can only be used on partly transparent materials) or be employed as a way of reducing complexity—since our brains are so limited. For instance, we often represent the behavior of an airplane by a projection of its three-dimensional flight path onto a two-dimensional map. Instead of

$$\text{path} = f(\text{latitude, longitude, altitude})$$

we project onto

$$\text{path} = g(\text{latitude, longitude})$$

The projected path may be thought of as the trace of the *shadow* of the airplane with the sun directly overhead, hence the use of the term "projection" for such a description. Naturally, many different flight paths will have the same shadow, or projection, since some information about the flight—namely, the altitude—is lost in this representation.

Taking a different projection, with the light shining from the front or the side, we would lose different information. But just because information is lost does not mean that the representation is useless. The projection from the side, which shows the altitudes throughout the flight, is useful for calculating fuel consumption in flight planning, since fuel consumption depends most critically on altitude. In this way, variables of no current interest may be "projected out" of the system, thereby reducing the computational load on the observer.

Of course, when information is lost due to projection, the observer who wishes to replace it may make an error. When we look at three-dimensional figures represented in two dimensions by projection, such as the famous Necker cube (Figure 6.7), we supply extra information about the world of objects to replace the lost information. The Necker cube is so perplexing because there is no clue as to which of two equally good interpretations (face with center dot in front or in back) to chose, and the figure spontaneously reverses itself if we stare at it long enough.

But if the Necker cube is really the projection of a wire figure hanging in space, these two interpretations are not the only ones, for there are an infinite number of wire figures that would give the same shadow. This infinitude is the *real* ambiguity in the projection—the illusion is in not suspecting that there is an illusion.

Figure 6.8 illustrates how we fool ourselves. Hans Elias[8] in writing about this figure says:

From our early student days we have become conditioned to identify the magnified image of an artificially stained slice with reality. During a scientific meeting . . . I introduced a talk on spatial interpretation of sections with a

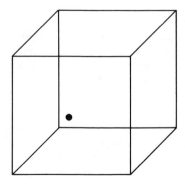

Figure 6.7. The Necker Cube—a projection of a three-dimensional figure onto two dimensions.

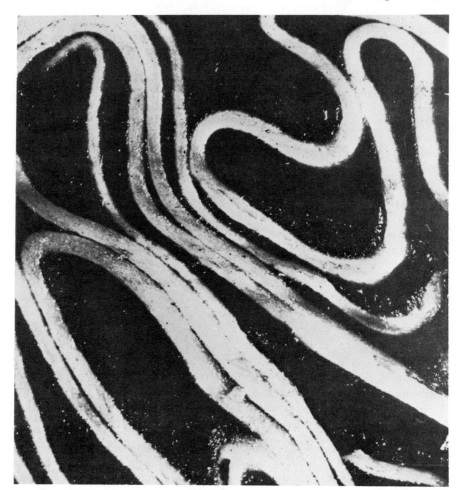

Figure 6.8.   What is it?

lantern slide, saying "Ladies and Gentlemen, I give you 30 seconds to identify the structure seen on the screen. As a clue, be informed that it is a section." Those who responded said unanimously, "It is a fiber." In reality this picture is a photograph of a section through a large, folded sheet of lasagna. . . .

Elias correctly attributes this response to scientific training:

Persons who look routinely at "sections" soon begin to identify the section with the real object, even if it is projected on the screen. Instead of saying, "this is a projected image of a stained slice showing an oblique section through the duct of a gland," we are accustomed to saying, "this *is* the duct of the gland."

If we are going to use metaphors from microscopy, then let us be sure to use metaphors from properly done microscopy. We can learn much from stereologists like Elias. We can learn, if nothing else, proper caution in our speech, which will inevitably lead to proper caution in our thought about subspaces. The lasagna picture, the Necker cube, the old–young woman—all carry the same lesson, which we may summarize as the *Picture Principle*:

When speaking about a dimensional reduction, insert the words "a picture of" in whatever you were about to say.

For example, if we say

Figure 6.7 is *a picture of* a cube.

we are less likely to get into trouble than if we say

Figure 6.7 is a cube.

At least we are reminded for a moment that there might be information that has been removed and that we might want to put back.

To put back the information lost through projection, we would have to have access to another source of data about the system, that is, about the missing dimensions. This inverse operation may be called *expansion,* and is one of the important reasons for the value of the state space point of view. Not that we are so often putting back lost information, more likely than not we are putting in information we never had. If we have been studying a system and discover that our view was incomplete, we do not have to throw away all our work. We simply add a new dimension for each newly discovered variable. Our old work is preserved because our old state space is now a projection of the new one, so all of our previous observations retain a meaningful interpretation.

For example, we may discover that temperature is important in describing the behavior of a system. This discovery may come only after we have done numerous experiments that happened to be at a constant temperature, so that we did not at first detect its importance. The descriptions of the system derived from these experiments now become descriptions of a particular projection of the system, a projection that eliminates temperature by holding it constant.

If we had not been controlling temperature, we could have seen "different" behaviors in this projection as the temperature varied unbeknown to us. In particular, we might have seen the line of behavior of the system crossing itself, which—if we believed the system to be state determined—would indicate something wrong with our point of view. Since a crossing represents two different paths emanating from the same point, two different successors to the same state, a crossed line of behavior cannot represent a state-determined system.

In a projection, on the other hand, crossing does not imply that the system is not state determined. An airplane spiraling down for a landing may never return to the same spot in (latitude, longitude, altitude), but its shadow (latitude, longitude) may cross and recross itself many times. This suggests a heuristic principle associated with behavior in the state space—the *Diachronic Principle*:

> *If a line of behavior crosses itself, then either:*
>   *1. the system is not state determined*
> *or*
>   *2. we are viewing a projection—an incomplete view.*

We may not know *where* to look—any more than a two-dimensional Flatlander knows where to look for the source of visitors from the third dimension[9]. At least, however, we know that we *should* look, which is more than half the battle.

When we employ the state space for *synchronic* representations, we obtain a static arrangement of points in the space at a particular moment in time. There being no "movement" among these points, we have no line of behavior to guide us. Instead we may develop a heuristic by analogy with physical space in which "no two objects can be at the same place at the same time." In an abstract state space this synchronic ("at the same time") rule need not apply, but it will if we have a *complete view* of the systems in question.

We may frame this heuristic device as the *Synchronic Principle*:

> *If two systems occupy the same position in the state space at the same time, then the space is underdimensioned, that is, the view is incomplete.*

For a complete view, every system must have its *unique* place, which is ultimately what we mean by "complete" and by "system."

In Figure 6.6, we notice that certain regions are occupied by more than one type of biome. At (60, 30), for example, there are coniferous forests, deciduous forests, and grasslands. Evidently, these two properties—mean temperature and mean precipitation—are not sufficient to partition the set of all biomes. Instead of

$$B = f(T, P)$$

where

B = biome type
T = mean annual temperature
P = mean annual precipitation

we have

$$B = g(T, P, \ldots)$$

with $f(T, P)$ being only a *projection* of $g(T, P, \ldots)$. What other factors might be involved? The state space can tell us no more: it could be minimal winter temperature, mean hours of sunlight, mean wind velocity, presence of man, or any or all of these. Drawing this projection is only the first step in a long and uncertain process.

Can the state space metaphor help us in dealing with the behavior of our OCCULT system with 10^{100} states? Since it is already partitioned into 100 members, we could readily use a 100-dimensional state space, but that may not prove helpful. Is there some way we can *reduce* the dimensionality of the representation?

Our first thought is to try projection. Projection would mean singling out a few of the members for special attention, which is actually what we did in Figure 6.3. But if only two or three members are chosen, most of the time they will not change caste, and when they do, the change will look quite mysterious and sudden.

But projection is not the only technique for reducing the dimensionality of a system. Recall that we originally decomposed the single line of behavior of a complex system by finding "properties" that then became "dimensions" in the state space. We can reverse the process and combine many properties into a few, preserving a little bit from each property instead of discarding some properties entirely as we do in projection. Any such process is called a *transformation* of the point of view; and although all projections are transformations, not all transformations are projections. These transformations, of course, are the metaphors applicable to the state-space way of looking at the world.

Just to take one example, we could reduce the OCCULT system to two dimensions by partitioning the members into two groups of 50 members each. Then we could reduce each group to a single number by adding all the caste values together. Since we are simulating the system on a computer, we can easily make this happen with a program segment that reads something like this:

4. $y = $ sum of $(d_1, \ldots, d_{50})$
5. $x = $ sum of $(d_{51}, \ldots, d_{100})$
6. Plot the point (x, y) on a graph.

We may then see a trajectory through the two-dimensional state space as shown in Figure 6.9. Although the arrows are not drawn, the general movement of this trajectory is down the page towards nihilism $(0, 0)$, which we could observe if we watched the computer plot the points one

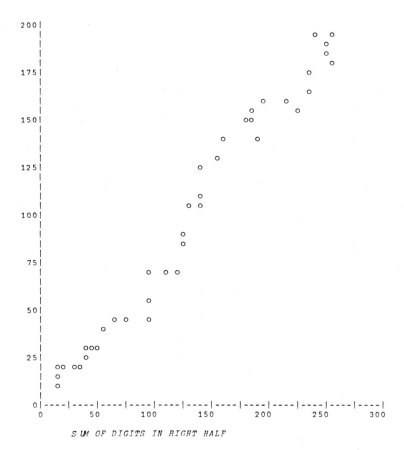

Figure 6.9. Trajectory of our OCCULT system through a "left–right" state space.

at a time. The OCCULT club, in this view, wends its way to nothingness, and then, it wends no more—just as Mr. O'Teric told us. We are thus encouraged that our simulation has been partially *validated,* for it seems to capture the behavior of its interpretation.

Although this transformation has thrown away much information, or, rather, *because* this transformation has thrown away much information, we learn something about the system's behavior that might not have been obvious in some other view. Even though we simulated the system and thus knew "everything" about it, we might not have realized that it would inevitably go to (0, 0), had we not been informed by the Grand Mystagogue.

What would the "white-box" analysis show? The representation of Figure 6.9 calls attention to the state (0, 0), where the sum of each half of 50 digits is 0. How can that be? Only if all digits, d_1 through d_{100}, are 0 will the two sums be 0. How did this nihilistic state come about, and why does it not change? Since the box is perfectly white, we can examine its structure for the source of this stationary state. We see that if j's caste is X (0), his new caste will always be X. Once a particular member becomes an X, there will be no member who can convert him to any other caste. This accounts for the state not changing from (0, 0).

Why does the state move *toward* (0, 0) in the first place? If an (i, j) is chosen for which i's caste is X $(d_i = 0)$, then t will be 0 and d_j will be set to 0, regardless of its previous value. Eventually, even a single nihilist will be picked as teacher by some (i, j) pair, and then there will be two nihilists. After that it will be twice as likely for this caste to be chosen as teacher, and a third X to be produced. In fact, we do not even need any nihilists to start the process, for $5 \times 2 = 10, 5 \times 4 = 20, 5 \times 6 = 30, 5 \times 8 = 40$ will all produce a zero starting with only a five and some even number. Thus, a nihilist will eventually appear and spread himself like the plague throughout the entire club, which explains why excluding nihilists did not save the OCCULT clubs.

Did you notice this property when we first built the white box? Perhaps you did, but many people do not. Just because we build a white box does not mean we will see all of its consequences. Once the property has "emerged," the white box makes it easy to uncover its "source"; but without observing the behavior, using this transformed view, we might never have seen the property at all.

Time as a Standard of Behavior

> The moving finger writes, and, having writ,
> Moves on; nor all your Piety nor Wit
>   Shall lure it back to cancel half a line
> Nor all your Tears wash out a word of it.
>   The Rubaiyat

One of the shortcomings of state-space representation is the deficiency in our brains that prevents us from visualizing n dimensions for n greater than two or three. But even worse is the deficiency of two- or three-dimensional space as a medium for communication. Though we might be able to work out n-dimensional problems in our heads, how do we communicate with each other about them in three-dimensional space?

Projections and other transformations can help us overcome this limitation, but all forms of the state space we have used so far have a further disadvantage. As we saw in Figure 6.9, we have no idea *how fast* a system is moving along a particular trajectory. Did it take 50 steps to get to (0, 0)? Or 50,000? Can we discriminate, on an airline map, between a 600 miles-per-hour 747 jet and a 100 miles-per-hour Piper Cub?

Paradoxically, one way to cope with the surplus of dimensions is by introducing another dimension, the dimension of *time*. Among all possible dimensions, time has the singular property of always moving in one direction. Time, in other words, cannot cycle. If, for example, you noticed a little calendar clock on Humpty Dumpty's music box, you could then expand your state description to read

$$S = (L, W, t)$$

Since t will never take on the same value twice, regardless of all your piety and wit, you entirely eliminate the possibility of a cycle or a crossing of any kind. Cycles are no longer the same states traversed repeatedly, but *similar* states being traversed *at different times*. Moreover, *measured* time enables us to discriminate between similar cycles that progress at different *rates*.

Jim Greenwood has pointed out to me that this concept of time as unidirectional is a construction adopted by physicists and others precisely for its unraveling properties. Other cultures, such as American Indians, often adopt a more cyclical view of time, a view that is not unfamiliar to our own culture:

> She was like all the old folk, she did everything in strict rotation. That is how they all thought and lived. It was always washing on a Monday and baking on a Wednesday. It could be raining cats and dogs on a Monday but she'd still wash—sheets, flannel shirts and all. Like as not, Tuesday would be hot and she would have burnt half the coal up Monday night getting it all dry. I've heard her say time and time again, "If I get out of my routine I'm finished!"[10]

As Shelley remarked, "Time is our consciousness of the succession of ideas in our mind."[11] More than that, our conception of time patterns our ideas, and different conceptions of time, used on different occasions, can be powerful tools for changing our points of view. The physicist's view of time as unidirectional and always independent can be traded for the "frequency-domain" view in which all phenomena are conceived of as compositions of cycles—not very far from the consciousness of the American Indian.[12] With suitable practice, the electrical engineer becomes mysteriously adept with frequency-domain equations and diagrams, and as much captivated by his characteristic modes of thought as the English mother from Akenfield by her weekly routine.

Time expansion of our view clears the way to a simple technique for representing multidimensional systems on two-dimensional paper or blackboards. In Figure 6.9, the state of the system was given by two variables,

$$x = \text{sum of digits in right half}$$

$$y = \text{sum of digits in left half}$$

so that we could write

$$S = f(x, y, \ldots)$$

Now, suppose we insert a time dimension, giving

$$S = f(x, y, t, \ldots)$$

If we project first x and then y out of the picture, we get two subspaces

$$Q = g(y, t, \ldots)$$
$$R = h(x, t, \ldots)$$

Since t appears in each, these subspaces can be placed alongside each other with corresponding times aligned, giving the view in Figure 6.10. Now the direction of the behavior is clear, the rate of behavior is clear, and we can even see that individual values of x and y may increase for a time. Most important, though, is the possibility these *chronological graphs* give of representing more than two variables at once.

Chronological graphs are found in a plethora of forms, because they are the most effective practical way of reducing multivariable behavior to tractable representation. Figure 3.2 showed three chronological graphs. An electroencephalograph (EEG) is a chronological graph of the electrical potential at a particular point in the brain of a living animal. A set of EEG's taken at different points in the brain gives us a more complete representation than any single one would do, without resorting to n-dimensional representations. Business indicators are conventionally presented in terms of chronological graphs—stock-market index, levels of retail inventories, wholesale-price index, gross national product, and so forth—which economists employ to gain a more comprehensive picture of the total economic system. The progress of an individual business may be followed through chronological graphs

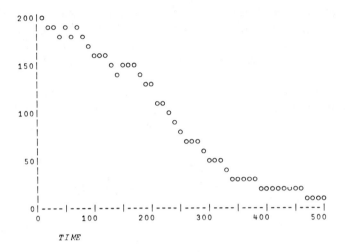

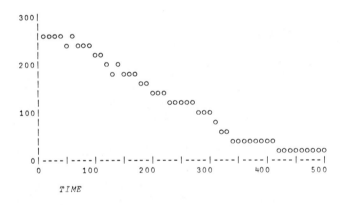

Figure 6.10.   Chronological graphs of the behavior in Figure 6.8.

of sales, inventories, production, and costs. To comprehend our complex weather, we chart the time-varying behavior of wind velocity, temperature, rainfall, barometric pressure, tides, glacial advance, ... But why continue? Anyone could extend this list without end.

The chronological graph is a tool of universal application. Paradoxically, one way to master the power of a tool is to probe its weaknesses. Thus, we offer the *Count-to-Three Principle*:

If you cannot think of three ways of abusing a tool, you do not understand how to use it.

Faithful adherence to this principle would protect us from the enthusiasm of the optimizers, maximizers, and other species of perfectionists—but mostly from ourselves. What, then, are some abuses of the chronological graph?

Look at the two graphs in Figure 6.11. The top graph represents what electrical engineers call, for obvious reasons, a "step function." The bottom graph is simply a slowly rising curve without much dramatic interest. Now look at Figure 6.12, where the two graphs are given their original labels. Which is the "step function" now?

Without knowing the *time scale*, talk of step functions and slowly rising curves is technical twaddle. Time scale has no meaning in an absolute sense, but only relative to some other time scale. While 4000 years might be a step function relative to the total evolution of man, 10^{-9} seconds might be a slowly rising function in a high-speed computer circuit. I recall with amusement that in 1957 we had a "zero-access"

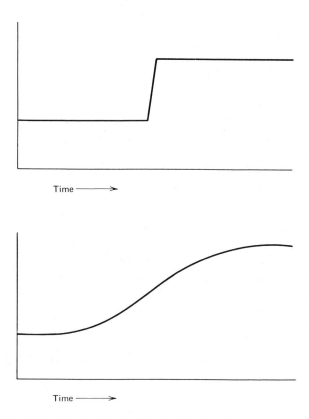

Figure 6.11. A "step function" and a "slowly rising function."

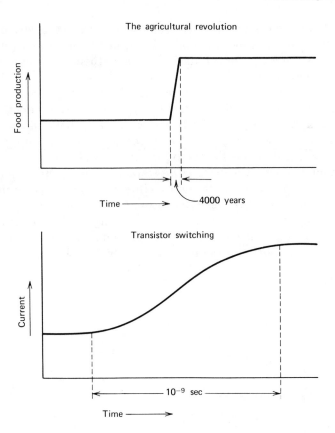

Figure 6.12. A slowly rising function and a step function.

memory installed on our computer, "zero access" meaning 96 millionths of a second. By 1967, we had "slow-speed" memory with 8 millionths of a second access time. Thus, twelve times faster than "zero" is "slow speed"—a decade later. When comparing the behaviors of two variables, then, time scales had better be the same. In Figure 6.10, the scales for *x* and *y* were different, which could lead to misinterpretation, but it is not terribly serious, since we do not know the "true" measure of either, or whether they should be comparable in magnitude. Time, however, is supposedly a *universal* standard.

Although we experience a "sense of time," this sense is hardly reliable enough for most scientific work. Therefore, we resort to standards—such as clocks—that provide reliable and disinterested scales to which other sequences can be compared. The absence of clocks hampered early work in physics to an extent which we can barely comprehend in this era of universally owned precision Swiss timepieces. Galileo would have given his left arm for such a watch.

In sciences that study a *historical* record—astronomy, climatology, ecology, archaeology, geology, paleontology, and the like—obtaining reliable "clocks" is as much a problem as it was for Galileo. For a while, radiocarbon dating was the great black hope for a wide range of dates, but with increasing sophistication has come knowledge that radiocarbon is not quite the uniform clock of our dreams.[13] As the variations in the carbon-14 pattern became evident, entire theories were dumped into the garbage heap of science—a lesson for all who ignore the problem of differing time scales and the Count-to-Three Principle.

Even when time scales are the same, the chronological graph may mislead us in more subtle ways. Every rose has a thorn. The very beauty of the time metaphor conceals its most dangerous trap. Time *separates* the variables on a standard scale, thus permitting us to handle systems with large numbers of variables. In so doing, it may lead us into unjustifiable feelings about the independence of these same variables. Since they *look* separated, we readily imagine them to be *independent*.

As we have seen, discovery of independent variables leads to economy of thought. Where there are dependencies, we could study fewer variables and obtain the same precision of prediction. Literally thousands of different EEG's may be taken from an animal brain. In striving for a workable analysis, we should like to eliminate all but a few EEG's—from which the others, if necessary, could always be inferred. But, by using chronological graphs to simplify our view, we may fail to notice dependencies that could simplify our view much more.

As usual, the choice of system properties must be a compromise between the *convenience of independence* and the *necessity for completeness*. Take our OCCULT system as a sample. Our first view (Figure 6.4) is certainly complete—we defined it that way—but it is difficult to use for extracting any patterns of behavior. Our partition into two summed halves (Figures 6.9 and 6.10) reduced the complexity of the system and permitted us to see a certain trend in its behavior. Did we throw away too much in this transformation? That depends, ultimately, on what we want to know.

Numerous other views might have been chosen. Suppose we had decided on a set of 10 variables, each variable being the number of members in the club having a particular caste. For instance, in the initial state of Figure 6.4, there are 4 zeros, 5 ones, 2 twos, 3 threes, 1 four, 9 fives, 6 sixes, 2 sevens, 3 eights, and 5 nines. (See Notational

Exercise 6.1.) This state would be represented as
$$(4, 5, 2, 3, 1, 9, 6, 2, 3, 5)$$
From this viewpoint, a rather different picture of system behavior emerges, which we can represent as a series of ten chronological graphs, as shown in Figure 6.13[14].

These 10 variables do not form an independent set, for we observe that once 9 of them have been determined, the 10th is always determined uniquely. Had we not simulated this system, but approached it from some higher level of observation, we would have *discovered* the nonindependence of the variables inductively. We might have noticed that as one variable went up, the others went down. After much labor, we might have wrought the law into a precise form—the form of a conservation law:

The sum of the variables of this system is a constant (100).

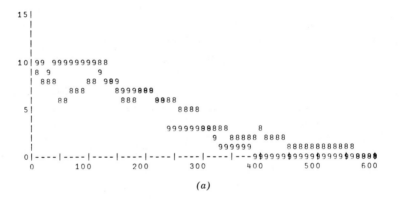

(a)

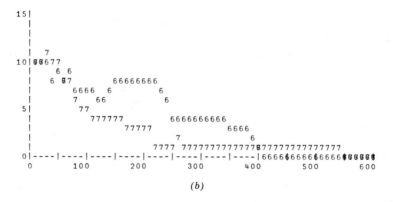

(b)

Figure 6.13.   Chronological graphs of OCCULT club membership.

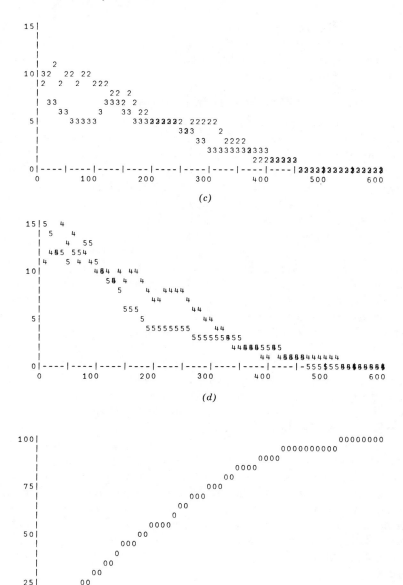

Figure 6.13. Continued

If we are inclined to put names to things, as some scientists and inventors are wont to do, we might say that our variables measure "endigitry." Then the law could be stated in more elegant form, the "First Law of Simpledynamics":

Endigitry can neither be created nor destroyed.

Do not be deceived by the elegance of this "First Law of Simpledynamics." It is not a general systems law, but a "special systems law," describing how our white box looks from this particular point of view. As before, we can take all the mystery out of this law by dropping to the white-box level, where all is "explained" if we but take the trouble to work it out. In this case, what our experimenter calls "endigitry" is no other than the number of digits in the state of the system, or the total number of members in the club. What he has discovered is that whereas the cells holding the castes in the computer's memory may change value, there are always 100 of them. While from the white-box point of view this may seem a law of the utmost triviality, from the black-box point of view it is a genuine discovery.

As before, the experimenter was able to discover this law because of the point of view taken. The law did not exist at all in the point of view of Figure 6.10, and would be very difficult indeed to determine from the point of view given there. Other points of view would give other insights, other special systems laws. For instance, suppose that instead of counting each caste separately, the even castes were all counted together and the odd castes were all counted together. (See Notational Exercise 6.2.) Some state such as (7, 15, 8, 12, 10, 8, 2, 19, 9, 10) in the previous view would give the state (36, 64) in this new system—since $7 + 8 + 10 + 2 + 9 = 36$ and $15 + 12 + 8 + 19 + 10 = 64$.

Given this view, the experimenter could hardly help but discover a new law by observing that the first number never decreased. While to us this fact is a simple consequence of the laws of multiplication (since if either member's caste is even, the new caste must be even), it again represents a genuine discovery by the experimenter. He might name the first variable "eventropy" and state the "Second Law of Simpledynamics":

Eventropy can never decrease.

This "law" was available to the observer working in the 10-dimensional state space, but would not have been quite so easy to discover in that view. The observer would first have to discover which state variables represent "eventropy." When he finally sees that 1, 3, 5, 7, and 9 should be added together, he will cry "Eureka," publish a paper, become famous, and win a Nobel Prize.

Science may be thought of as the process of learning which ways of looking at things yield invariant laws. The laws of science may thus be *descriptions* of how the world looks ("Eureka"—I have found), or *prescriptions* for how to look at the world ("heuristic"—how to find). We really have no way of knowing which.

Behavior in Open Systems

The second principle (of thermodynamics) means *death by confinement....* Life is constantly menaced by this sentence to death. The only way to avoid it is to prevent confinement.... Confinement implies the existence of perfect walls, which are necessary in order to build an ideal enclosure. But there are some very important questions about the problem of the existence of perfect walls. Do we really know any way to build a wall that could not let any radiation in or out? This is theoretically impossible; practically, however, it can be done and is easily accomplished in physical or chemical laboratories.

L. Brillouin[15]

Why do physical and chemical laboratories try to build ideal enclosures? To create state-determined systems for study. And why do they like to study state-determined systems? Because in a state-determined system the behavior is simple. Everything that the system can do expresses itself in a single uncrossed line of behavior.

The observer, of course, can introduce differences. He might happen to see different behaviors by approaching such a system at different times, so that he would see different parts of the line. Though each day repeats its brother, we do not see the dawn if we wake at noon. A second observer might see different behaviors because he is lumping the system in a different way, or partitioning different properties, or using a different time scale. Even a single observer might be "different" at different times, for there is no reason why he cannot *change* his lumping, partitioning, or time scale.

But if the observer takes care of all these matters and also succeeds in isolating his system within perfect walls, the line of behavior may still be tangled, in which case he says he sees "randomness." The observer, however, has no reliable way to distinguish randomness from concealed openness— or "leaky walls." There may be an occasional high-energy cosmic ray smashing in at irregular intervals, or elves may be sneaking in at night and fiddling with his apparatus.

The state-determined system was characterized by a relation

$$S_{t+1} = F(S_t)$$

perhaps with S_t suitably defined to encompass past behavior, according to the Eye–Brain Law. The "random" system is characterized by

$$S_{t+1} = F(S_t, \ldots)$$

where the "something else" is unknown to the observer. We might just as well say that

$$S_{t+1} = F(S_t, R_t)$$

giving the name R to that random "something else." But this is exactly the same form as we would use to describe an *open* system:

$$S_{t+1} = F(S_t, I_t)$$

where I is the "input," something that comes from "outside" the system. By the Principle of Indifference we know it should not matter whether we call "something" I or R, so we shall assume, for now, that R "comes from outside." By the Principle of Difference, however, we know that this choice of wording will make some people unhappy, but they will have to hold their tongues.

From this point of view, any closed system is state determined. Every line of behavior in a finite state-determined system must *eventually* terminate in a cycle of one or more states. Why? Because if there is a *finite* number of states, eventually one state— call it S_x—has to be reached a second time. It will then be followed by S_{x+1}, which we know from

$$S_{x+1} = F(S_x)$$

will be the same state that followed S_x the previous time. In a similar way, S_{x+2}, S_{x+3}, and so on, must be forever the same, which is the definition of a cycle.

Cycles are the signature of state-determined behavior. When we see a system in a cycle, we suspect that it may be at present uninfluenced by external factors. Of course, it may be influenced by a *cyclic external factor,* or by an external factor that is *too small* to break the cycle. The repeating dreams or thoughts we get when "lying awake with a dismal headache and repose is taboo'd by anxiety" suggest that we do not have sufficient external input to come out of the cycle. We cannot even break out consciously, because the thought needed to break out is not part of the cycle. Only when a car screeches its brakes or a tomcat knocks over the garbage-can lid do we get strong enough input to put us out of our state-determined misery by waking us up entirely.

The closed-system fiction is thus a useful heuristic device. If we see noncyclic behavior, we look for an input. If, on the other hand, we asserted that the system is closed, but "random," we would be saying that there is *no use looking* for any additional input. Many scientists are reluctant to admit that a system is open, so sometimes it does pay to talk about randomness as a labor-saving, or face-saving, device. That way we do not have to admit that our view is incomplete or explain why we are not looking for inputs.

If a state-determined system is partitioned into a "system" and "environment," the "system" part will, in general, no longer be state determined. For example, when you made the acquaintance of your first music box, you thought its state was given by

$$S = (R, G, W)$$

which was true as long as you didn't kick it. The view seemed complete enough to be state determined, for you saw a cycle.

Paradoxically, we seek state determinedness with all the fervor of a knight wooing his distant lady, but, should we ever obtain the object of our burning desires, like the knight we immediately lose all interest. State determined is too "perfect" to be interesting, so you kicked the box, transforming its state description to

$$U = (R, G, W, \text{kick})$$

By the Principle of Indifference, you may either see this new state or retain your old one by writing

$$S = (R, G, W)$$
$$S_{t+1} = F(S_t, \text{kick}_t)$$

Between prods, the value of "kick" is zero or some appropriately innocuous symbol. From this point of view, the kick is seen to have a *selective* role in determining the next state, and thus the next cycle.

Because of our love of simplicity, we tend to think of systems as having a single line of behavior, which we simply call "the behavior of the system." An open system, however, has no single line of behavior, but a repertoire, or set, of behaviors selected by the input. Therefore, we can speak with less certainty about open systems.

To speak about open systems, we have to renounce speaking about *the* behavior of the system, though we might want to use Behavior with a capital "B" to permit us to speak of the entire repertoire. For example, when Father interrupts Johnny's finger painting on the living room wall and says, "I don't like your behavior, young man," he is using the word in its first, more specific, meaning. When Johnny's teacher gives him an F in Behavior, she is speaking of the entire set—or at least that part of it he displays in response to school inputs.

In closed systems, the paradigm of a law is

If the system is closed, then the behavior is. . . .

In open systems, however, the paradigm is

If the input is thus and so, then the behavior is. . . .

We generalize our open system laws by somehow characterizing the set of behaviors in response to a certain *set* of inputs, as in

If the input is any one of the following . . . , then the behavior is. . . .

or

If the input is any one of the following . . . , then the behavior is in the set. . . .

For example, if I want a 25-cent candy bar from a vending machine, I can insert a quarter, two dimes and a nickel, one dime and three nickels, or five nickels. Any input in this set gives the same output—a candy bar. "Inside" the machine is some mechanism that we could simulate on a computer using a *conditional* operation such as

1. T = total coins inserted since last candy bar.
2. If T = \$0.25, then give candy bar. Otherwise wait for more money.

Machines with more complex repertoires of behavior will of necessity have more complex programs for determining what to do for each input case. For example:

1. T = total coins inserted since last candy bar.
2. If T is greater than \$0.25, then make change and reduce value of T to \$0.25.
3. If there is not enough change, then
    return money,
    set T to zero,
    and turn on NO CHANGE light.
4. If T = \$0.25, then release selector bar and wait for selection.
5. If selection is licorice, then give licorice.
6. If licorice is empty, then return money. . . .
7. . . .

Clearly, such a program could be continued indefinitely to produce a machine with arbitrarily complex behavior.

Our OCCULT system is also an open system, with the selection being performed by the input pair (i, j), according to the simulation program. Therefore, what we see when observing its Behavior will depend on the sequence of inputs that happens to arrive during our period of observation. We expect, therefore, that if we simulate this system many times, starting from the same initial state, we might see any number of different behaviors, even though we do not change our observational characteristics.

If the behaviors we see are noticeably different, we shall be upset and feel we do not "understand" the system—just as Johnny's schoolteacher does not understand why he sits quietly all year and then on one Friday afternoon sets fire to the library. She will be at a loss to characterize this "behavior," and will probably settle for an F, even though he has been a model pupil virtually all year.

Her decision to characterize his Behavior according to one isolated instance of behavior is one way we simplify Behavior of open systems. Sometimes we characterize a person by the behavior he displays *most* of the time—like a "professor," "gourmet," or "golfer." In Johnny's case, however, we would tend to pick one exceptional act as standing for the entire set.

In common speech, we apply the name "liar" to anyone who tells a single lie—yet we have no word for someone who invariably tells the truth. A multitude of other words, all implying badness, are applied for a single act of the denoted Behavior: murderer, embezzler, loser, cheater, sinner, adulterer, and drunkard. The existence and use of these words indicates a particular yearning for determinate behavior.

If a man murders his wife, he is incarcerated for life, so he will not repeat the act on some other innocent victim. The penal code reifies the concept that his behavior is state determined, the belief that next time around the cycle, this "murderer" will murder again. But the modern psychiatric view says that he, of all people, is unlikely to murder again—that murder is determined by the environment, by the one last nag that snapped the milksop's thirty years of control. Of course, there are other Behaviors all lumped under the title of "murder," so that this determinate view is not always so cruel and inappropriate. It is for these other murderers, the "psychopathic killers," that our laws and conceptions seem to be formed.

Certainly there are many systems for which it seems proper and necessary to characterize the entire set of behaviors by one particular line. What would we think of an engineer who had just designed a bridge that was "safe" because "it won't fall down more than once in five years?" In other words, whatever else may interest us in the Behavior of a system, we usually want to know about the chances of it ever displaying a previously unobserved and sufficiently disastrous behavior. No doubt we could make bridges for a fraction of the cost if we were willing to let them fall down every few years; but, usually, a major share of a bridge's cost is devoted to making such chances sufficiently low.

Because of our fear of the unexpected, we usually observe a system for a period of time before we are willing to make statements about its complete repertoire of behaviors—only the very young think a "pickup" is a proper introduction. The time we take to make these observations depends on a number of factors, but particularly on our expectations, which are based on experience with "similar" systems. Were we to receive a box that emitted a loud, ticking sound, we might suspect, no matter how long we observed the ticking, that some unique form of behavior was imminent. If the box contained a bomb, no amount of observation would tell us when and if it would explode; for that is precisely the idea behind a time bomb—that we should be innocent of its capability for discontinuous behavior.

Characterization by typical behavior and by exceptional-but-important behavior are two ways we conventionally attempt to recover the single line of behavior we enjoyed in closed systems. Another technique is to take *average behavior,* either a diachronic average—a "hot summer" or a "wet winter"—or a synchronic average—a "nervous breed" or an "unreliable brand." In fact, to transfigure open-system behavior to a determinate form, we may employ just about any technique of abstraction. If we still do not succeed, we have a final trick up our sleeve: we characterize the behavior as "random," "adaptable," "unpredictable," "crazy," or "eccentric" (which means "crazy but rich"). In this way, we can *always* succeed in reducing Behavior to a single behavior.

Moreover, in our quest for reduction of the Behavior set, we need not remain impotent observers. Instead of *imagining* the system to have a single behavior, we can *ensure* that it does. In the words of the gangland chief, we can "make it an offer it can't refuse." Though we may be unable to determine whether a ticking box contains a bomb, we can simply dunk the box into a pail of water. Dunking the box protects us *only* because the bomb is to some extent an open system. It can exhibit one of several behaviors, depending on its environment. Since we may act *either as observer or as environment,* we can either *predict* its behavior or *influence* its behavior. But even here, the two roles are not separable, since we have no way of knowing that the box does not contain a bomb triggered by wetness:

1. If moisture level is too great, then close the switch!

Open systems baffle us, and we prefer to think of our systems—or create our systems—to be as closed as possible. Openness is a puzzle because it complicates prediction and observation, yet at the same time, it lets us gain predictability by acting on the system. Before pursuing this paradox, let us consider, by looking at the OCCULT system, the way the system's behavior is influenced by its initial state (the state-determined part) and its input sequence (the open-system part).

The Principle of Indeterminability

Don't you know what our holy Scriptures say? The Law was written nine hundred and fourteen generations before God built the world. But it wasn't written on parchment, because at that time no animals existed to give up their hides; nor on wood, for there were no trees; nor upon stone: there still were no stones. It was written in black flames upon white fires on the left arm of the Lord. It was within this sacred Law, I want you to know, that God created the world.

Nikos Kazantzakis[16]

Although our OCCULT club is not quite the world, it too must start somewhere. How a club starts is determined by something outside our simulation, that is, the original 100 members of a club and their castes are chosen before the simulation begins. However they are chosen, the starting state will have some influence on what happens later. If the input is random, most possible starting states, such as (13, 12, 8, 15, 7, 16, 5, 4, 8, 12), will show a behavior much like that in Figure 6.13. Figure 6.14, for instance, shows the behavior starting from an entirely different state with another random input sequence, yet it is hardly distinguishable from Figure 6.13. There does indeed seem to be some "Law" operating to bring about this nihilistic state of all zeros: (100, 0, 0, 0, 0, 0, 0, 0, 0, 0).

A system that has the property of reaching the same final state almost regardless of the initial state and input sequence is called an "equifinal" system. Equifinal systems appeal to our need for consistency in behavior and simple characterization of what we observe. Of

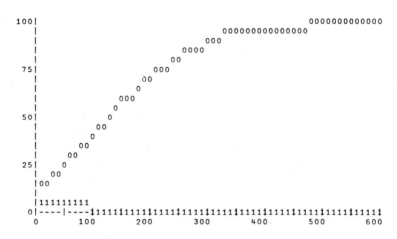

Figure 6.14. Chronological graphs of OCCULT club membership.

course, sometimes we are interested in *how* the system gets to its equifinal state, which may not be the same in all circumstances. The paths of glory lead but to the grave, but they are different paths, and that is what makes life so interesting. Nevertheless, we are often attracted to the property of equifinality because it indicates some kind of structure in the process by which the input is allowed to act on the system.

Be sure to notice, though, that we said "almost regardless" in our definition of equifinality. We can certainly prevent an OCCULT club from reaching the equifinal state by starting it in the state (0, 50, 0, 0, 0, 0, 50, 0, 0, 0), that is, all ones and sixes. In that case, it will be driven by random input pairs to the state (0, 0, 0, 0, 0, 0, 100, 0, 0, 0) since only ones and sixes can result from ones and sixes through the step:

$$2. \ t = d_i \times d_j$$

($1 \times 1 = 1$; $1 \times 6 = 6$; $6 \times 1 = 6$; and $6 \times 6 = 36$, which gives 6 when the last digit is taken.) Such behavior is shown in Figure 6.15, where only the sixes and ones are plotted since all other digits remain at zero. Eventually, the sixes "drive out" the ones.

Or, consider what happens when there are only odd castes to begin with—ones, threes, fives, sevens, and nines, as in the state (0, 20, 0, 20, 0, 20, 0, 20, 0, 20). Without any zeros or even castes around, the fives take on the dominant role that the zeros played before, and the state will eventually become all fives—(0, 0, 0, 0, 0, 100, 0, 0, 0)—as shown in Figure 6.16.

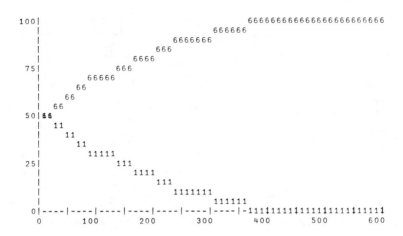

Figure 6.15. Chronological graph of OCCULT club membership (*SV*).

By studying the white-box structure, we can ascertain that there is one more such equifinal state. Any time there are only ones left, the final state must remain fixed, since $1 \times 1 = 1$. Moreover, since $9 \times 1 = 9$, and $9 \times 9 = 81$ (which yields 1 in our algorithm), once we have only nines and ones left no other digits will ever be introduced. This region of only nines and ones can be reached from the region of only ones, threes, sevens, and nines, since these four digits together can only produce themselves. So we can see that an initial state with no even numbers or fives will eventually lead to a state with only ones and nines, which will in turn lead to a state with only ones, which should make Mr. O'Teric jubilant.

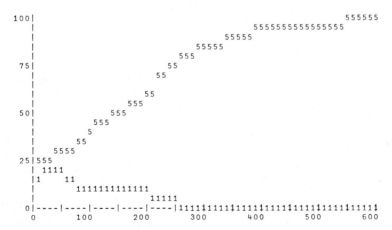

Figure 6.16. Chronological graph of OCCULT club membership (*SVI*).

Symbol	Equifinal state	Starting membership
SI	(0,100,0,0,0,0,0,0,0,0)	No members of caste V, X, II, IV, VI, VIII
SV	(0,0,0,0,0,100,0,0,0,0)	At least one member of caste V and no members of caste II, IV, VI, VIII, X
SVI	(0,0,0,0,0,0,100,0,0,0)	No member of caste X or V and at least one member of caste II, IV, VI, VIII
SX	(100,0,0,0,0,0,0,0,0,0)	At least one member in caste X or At least one member in caste V and one in caste II, IV, VI, VIII

Figure 6.17.   Four "species" of OCCULT clubs.

In summary, then, an OCCULT club has four different equifinal states at which it will "stick" without change once it gets there, as shown in Figure 6.17. Notice that the relation "leads to the same equifinal state as" satisfies the reflexive, symmetric, and transitive conditions. Thus, we can partition the state space of our system into four distinct regions, much as Figure 6.6 (almost) partitioned the (temperature, precipitation) state space into six biomes. We might name these regions after their equifinal states—*SX, SI, SV, SVI.*

If we did not know the internal structure of this white box, and if we observed many clubs each started with its own peculiar input state, we might say that there were four "kinds," or "species" of clubs. The species of a club is determined by its initial state, but we do not know which type of club it is, except by seeing it go to its final state. In other words, we know a robin's egg from a lizard's egg because one eventually produces a robin and the other a lizard. It was all determined before the eggs were laid, but we may not know until they are hatched. Eventually, we learn to recognize other "characteristics" of each species, so we do not have to wait until the final outcome to make our identification.

How many clubs of each species will we see? Which will be the monad (*SI*) and which the quincunx (*SV*)? Translated into state-space terms, what is the chance of any club having an initial membership in each region? That we cannot say, for we are not sitting at the right hand of the Grand Mystagogue when the initial membership is chosen. If it were chosen *at random*, then the probability would be about 0.9999999999 that a box would be an SX, about 10^{-10} of being an SVI, about 10^{-30} an SV, and about 10^{-40} an SI. Thus, if Mr. O'Teric chose at random, we should never see any clubs other than nihilists (SX), and would be content to say that nihilism was "the behavior" of the system. But Mr. O'Teric is the Grand Mystagogue and may choose members as he likes. We might see any of the types, and our image of the system would then be quite different. Such is the possible influence of the system's initial state.

To study the influence of the input sequence, let us for the moment restrict our attention to the nihilist type, for which random input will eventually drive each of the members to caste X. Without asking the embarrassing question of exactly what Mr. O'Teric meant by random input, consider what sorts of behavior might be exhibited if the input were *not* random. For the first case, suppose that the input is constrained so that the pupil chosen is always the same, while the teacher is chosen randomly. Since only the caste of the pupil changes, only one member can ever change caste, so the various states through which the club passes can never be very far from the initial state. Starting from the same state as Figure 6.13, the club would never reach SX, even though the initial state is in that region. A robin's egg will not hatch if it falls out of the nest, and a club will not turn nihilist if Mr. O'Teric can control the selection of teachers and pupils.

Other nonrandom inputs—not so obvious—could also prevent the system from reaching SX. For instance, in one of the computer simulations, the mechanism used for generating a sequence of "random" inputs happened to exclude a few numbers from ever appearing as j. The members in *those* positions never changed caste, so that SX was *approached,* but not quite attained. Indeed, the very failure of the system to reach SX revealed the nonrandomness of the input mechanism.

Much more subtle nonrandomness can have a similar effect, either by prohibiting certain pairs in the input or by prohibiting certain *sequences.* Another experience with a computer simulation yielded a sequence that, though it permitted all 100 members to be either teacher or pupil, did not yield all possible member pairs. As it happened in this case, the numbers 12, 13, 37, 82, and 94 never appeared as pupil unless one of the others appeared as teacher. In the first random starting state, member 13 was a V and one of the others had an even caste, so all fives eventually degenerated to X. Similar behaviors were exhibited for subsequent initial states. Eventually, however, one initial membership was generated in which none of the five members of the clique was either X or V. After a very long simulation, they resisted nihilism, much to our surprise—until we discovered the nonrandomness of the input mechanism.

An observer viewing this club without complete knowledge of the system could arrive at one of two equally reasonable conclusions. Either the input to the system was nonrandom or there was a clique consisting of five members that somehow "resisted" the action of the input, or, at least, behaved differently from the other club members. We could, in fact, create a simulation that arbitrarily prevented members (12, 13, 37, 82, 94) from becoming pupils. We would need a statement in our program like this:

$$3.\ \textit{Unless } j \text{ is one of } (12, 13, 37, 82, 94),$$
$$d_j = \text{last digit of } t.$$

Such a simulation would behave in essentially the same manner as the undifferentiated system under the peculiarly partitioned input.

From considerations such as these, we may derive an important principle that rests only upon the most general grounds—the *Principle of Indeterminability*:

We cannot with certainty attribute observed constraint either to system or environment.

In specific cases, we might do worse than the principle permits, for the observer himself may be contributing the constraint. Eddington provided the classical example of observer constraint by describing the crew of an imaginary oceanographic vessel, which, by classifying the specimens snared by its nets, concluded that there were no creatures in the ocean less than 3 inches long.

The observer, of course, is part of the environment, so we should not be surprised to find him encompassed by the Principle of Indeterminability. Many instances of "3-inch nets" may be found in the history of science. The history of medicine, which deals with complex systems in a way that confuses the role of observer (diagnostician) and environment (clinician), is particularly full of fine examples.

Consider the case of Alexander Wood, who invented the first hypodermic needle around 1855, using it for the subcutaneous injection of a morphine salt to relieve local neuralgia. He had great success with the method, but he believed that the pain was only to be relieved by injections at or near the place of pain. Since morphine injected *anywhere* will give general relief of pain, he was successful at relieving pain even though he thus restricted his injections. But when faced with a woman with pain in her scalp, he complained of being unable to make an injection at this site and left the poor woman unrelieved. Because of his theory, it just never occurred to him to make the injection elsewhere.

It remained for Charles Hunter, in 1858, to discover that equal relief was given by injecting morphine in unaffected parts. Most probably, Hunter was freer than Wood to make this discovery, since he was not burdened (yet) with the pride of invention. It was Hunter, however, who invented the name "hypodermic," and he too soon succumbed to the parental pride that clouds the vision and prevents the child's development.[17] Now, of course, we know that "local injection" is a worthless idea—or do we? Has it ever been really tested?

The number of untested assumptions in science is staggering. On any day we can open any one of dozens of newly arrived journals and find reports of "discoveries" that were made simply by relaxing a constraint that was in the observation process itself. Here is an example from *Science*[18]. First, the abstract of the findings:

The so-called pony fish of the tropical and subtropical Indo-Pacific region can emit light from a broad area of its ventral surface. An experimental analysis of this luminescent system supports the hypothesis that it functions by emitting light during the daytime, which matches the background light and thereby obscures the silhouette of the animal.

A neat adaptive mechanism, but why wasn't it discovered before? After all, there are few things as spectacular as a luminescent fish. The author explains:

Since observations on bioluminescence in the field are commonly made at night, and in the laboratory almost always in a darkened room, it is not surprising that the possibility of bioluminescence by day has escaped interest and attention.

So things have not changed all that much from 1855 to 1971, for the Principle of Indeterminability still seems to hold. I cannot resist one more example from my own backyard—from the world of computers. It is not only scientific laws that are plagued by misplaced constraint. In the computer business, the Principle of Indeterminability asserts itself thousands of times a day, all over the world, as the following example will demonstrate.

We started this chapter with a glowing tribute to the simulation powers of the digital computer, but we would be remiss if we failed to point out that computing is not always as simple as we have painted it. When we write a program—to simulate our OCCULT system, for example—it does not always do what we thought it was going to do, usually because we have written an incorrect program. On the other hand, from time to time the very machinery of the computer does not do what it is told. When this happens, when "it is not operating properly," we have a "machine error," as opposed to a "programming error."

The machine is the environment in which the program is run, and sometimes an error occurs for which it is essentially impossible to say whether it is in the machine or in the program. Because machine errors are now so rare, the usual prognosis is that the error will be found in the program. When that prognosis is wrong, it leads to no end of wasted effort groping for the error in the program.

In the present example, two programmers had been laboring for weeks trying to find the source of an intermittent failure, when they became convinced that the error was in the machine. The engineers were called, and after a dispute over the source of the error, the engineers agreed to try to find it in the hardware. They pulled the operator's desk away from the side of the machine so they could gaze upon its vital organs. Opening *this* black box is not forbidden to engineers, but even by looking inside they could find nothing.

They gathered up their diagnostic programs and announced that the error must be in the programs, since the engineers' programs worked perfectly. After they left, the programmers tried their own programs again and failed, from which they inferred that the error must indeed be theirs.

Alas, this entire cycle had to be repeated before the programmers finally became reconvinced that there was a machine error. Again the engineers pulled out the desk, opened the machine, and went to work. Again, they found nothing. Three more times the entire episode was reenacted, always with the same results. And then, late one night, the programmers found that their program worked.

What had changed? The time of day? Surely that had no effect. Or did it? As they stood and watched their program grinding on its merry way, much like one of our music boxes, the janitor who was cleaning the floor finished his work and pushed the desk up against the side of the computer. Immediately, the program stopped!

QUESTIONS FOR FURTHER RESEARCH

1. *Natural Ecosystems*

Figure 6.18 is adapted from an article[19] on world patterns of plant energetics. It represents a three-dimensional state space rendered in

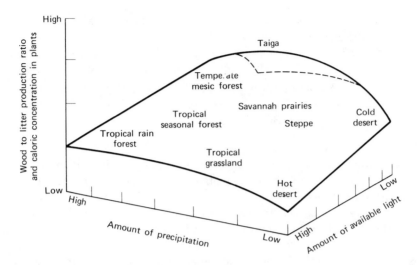

Figure 6.18. A three-dimensional state space related to the two-dimensional view of Figure 6.6.[16]

two dimensions, encompassing some of the same information seen in Figure 6.6. Discuss the relationship between these points of view, and search for other state-space diagrams viewing the same data with other decompositions.

2. *Data Reduction Techniques*

Of all the forms of lines of behavior, the straight line through the state space is in some sense the simplest. Mathematically, *any* continuous curve can be combed out if we work hard enough at it. Empirical points in a state space, however, are not really points, but regions whose size is determined by the range of observational "error." Each transformation has a characteristic way of expanding or contracting these error regions, as well as reshaping the "line." For instance, someone has suggested the following law, the *Log–Log Law*:

Any set of data points forms a straight line if plotted on log–log paper.

Discuss this law, and the behavior of other transformations in producing straight lines through two-dimensional state space.

3. *Componential Analysis*

One of the controversial methods of anthropological linguistics is based on the attempt to discern "native categories of thought" through decomposition into fundamental "components." Trace the development of this method and the controversy surrounding it in terms of the concepts of this chapter, and argue one side of the controversy or the other—or against both.

References: A. F. C. Wallace and J. Atkins, "The Meaning of Kinship Terms." *American Anthropologist,* **62,** 58 (1960).

Robins Burling, "Cognition and Componential Analysis: God's Truth or Hocus Pocus?" *American Anthropologist,* **66,** 20 (1964).

4. *Factor Analysis in Psychology*

The human mind seems to represent, for us, the ultimate black box—a box from which psychologists have been trying for three generations, at least, to extract the "true" dimensions of intellect. In fact, the study of intelligence was the spur that led Spearman in 1904 to develop a mathematical technique for automatically extracting "factors" or "components" or "dimensions," the technique we call "factor analysis." The method was much refined by Thurstone, and yet, even though it has been applied and refined by hundreds of workers over 70 years, the debate still rages over the "reality" of the factors it extracts.

Though the method is now applied far beyond the boundaries of psychology, the main focus of application still remains on the original question: What are the factors of intelligence? J. B. Guilford, who participated in much of this span of development, has summarized the current picture of intelligence obtained through factor analysis methods into a three-dimensional state space:

(Contents, Products, Operations)

which generates $4 \times 6 \times 5 = 120$ states, as shown in Figure 6.19. Discuss the methods by which this state space was derived, how it is used by psychologists, and in what ways it may prove inadequate to the task of comprising human intellect.

References: J. P. Guilford, *The Nature of Human Intellect.* New York: McGraw-Hill, 1967.

H. H. Harman, *Modern Factor Analysis.* Chicago: University of Chicago Press, 1967.

5. *Art History*

Over the centuries, what devices have been used by artists to remove the ambiguities inherent in projecting three-dimensional scenes onto two-dimensional canvases? How do the traditions differ in Occidental and Oriental art? What modern technologies may be involved in the next round of devices?

Reference: E. H. Gombrich, *Art and Illusion.* New York: Pantheon, 1960.

6. *General*

Find a business magazine and an engineering magazine and see how many chronological graphs you can find in each. Also see how many other forms of state-space representation you can find. In the systems described, identify the initial state and the input sequence, and discuss the effects that different ones might have on the behaviors displayed. If possible, describe each behavior in terms of some other representation than the one given.

7. *Film Making*

Animation on film gives us the possibility of representing time or other dimensions more effectively. Write a film script of an animated

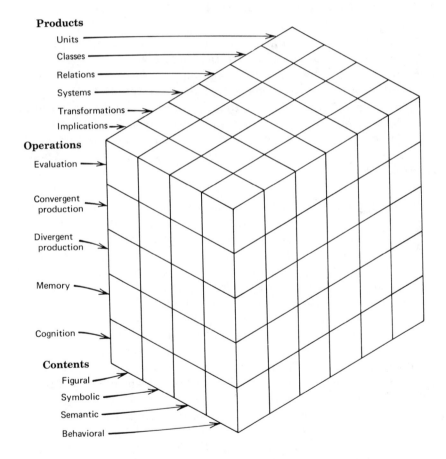

Figure 6.19. Guilford's three-dimensional model of intellectual activities.

film that could be used to put across the behavior concepts of this chapter, using at least three new views of system behavior suitable for the film medium.

8. *History*

Carbon-14 dating did not turn out to be the universal clock it was first hoped to be, but there are many other clocks that record time in the past, particularly the historical past. Suppose you wanted to study the weather changes over the last 1000 years. What kinds of clocks might you use? What distortions would be peculiar to each? After exhausting your ideas, have a look at

Emmanuel Le Roy Ladurie, *Times of Feast, Times of Famine: A History of Climate since the Year 1000.* Translated from the French by Barbara Bray. New York: Doubleday, 1971.

9. *Commuting*

Why is it that when you arrive at a bus stop, you are more likely to see a bus going in the opposite direction before your bus arrives? Because the bus is on a cycle, and you are more likely to be closer to your origin end than your destination—otherwise you wouldn't be using the bus. There is no reason to believe that the two parts of the cycle will be the same size, but you lump them into just two states. Quantify this observational artifact, and generalize it to more than one bus on the route. Also, analyze the folk wisdom that "a watched pot never boils" in a similar vein.

10. *Teratology*

Empirical observations show that women, as they get older, are more likely to have mongoloid (Down's Syndrome) children than are younger women. For instance, a pregnant woman at 40 is 10 times as likely to have such a child as a pregnant woman of 25. One explanation of this effect is that the ovum can be fertilized over a span of several days, and only if it is fertilized toward the very end of this span when a separation of chromosomes called nondisjunction has occurred, will it result in a mongoloid child. The frequency of intercourse in such a situation is analogous to the frequency of observation: with intercourse occurring daily, there is essentially no chance of fertilization occuring at the end of the span. But if intercourse occurs, roughly, every ten days, then the chance of such fertilization rises tremendously. Make a model of this phenomenon, relating the fertility cycle, the frequency of intercourse, and the frequency of mongolism. What sorts of empirical data would tend to corroborate this model?

Reference: Abraham M. Lilienfeld, *Epidemiology of Mongolism.* Baltimore, MD: Johns Hopkins Press, 1969.

11. *Microwave Engineering*

The introduction of increasing numbers and varieties of microwave apparatus has led to stories and speculation about interaction of microwaves with other systems. For example, we ordinarily think of kitchen appliances as well-isolated components of the kitchen, but when microwave ovens were introduced, it soon became clear that the presence of a microwave oven could have serious repercussions on someone who happened to have a heart pacemaker. To paraphrase an old saying, "If you can't stand the microwaves, stay out of the kitchen."

Discuss other situations in which microwave systems could prove interactive with previously well-isolated systems, and project the possible effects and actions as a result of those effects.

Reference: *IEEE Transactions on Microwave Theory and Techniques.* Especially "Special Issue on Biological Effects of Microwaves." **MTT-19,** 128 (February 1971).

12. *Psychiatry*

Discuss the meaning of the following paragraph:

The dynamic viewpoint applied in this book to social–psychological phenomena is fundamentally different from the descriptive behaviorist approach in most social-science research. From the dynamic standpoint, we are not primarily interested in knowing what a person thinks or says or how he behaves *now*. We are interested in his character structure—that is, in the semipermanent structure of his energies, in the directions in which they are channeled, and in the intensity with which they flow. If we know the driving forces motivating behavior, not only do we understand present behavior but we can also make reasonable assumptions about how a person is likely to act under changed circumstances. In the dynamic view, surprising "changes" in a person's thought or behavior are changes which mostly could have been foreseen, given the knowledge of his character structure.

Eric Fromm[17]

In particular, discuss the use of the word "behavior" in its various forms and occurrences, as well as the word "structure." What does Fromm mean by "changed circumstances," and how does the Principle of Indeterminability apply?

13. *Dimensional Analysis*

From dimensional analysis we get the idea of a *dimensionless product*—an expression involving two or more physical variables raised to powers in such a way that all dimensions cancel one another out. Examples of important dimensionless products are: Reynolds' number, Froude's number, Weber's number, and Mach's number—the last being simply the ratio of two velocities and therefore clearly dimensionless. "A set of dimensionless products of given variables is said to be complete, if each product in the set is independent of the others, and every other dimensionless product of the variables is a product of powers of dimensionless products in the set."

Discuss the significance of dimensionless products, and of the requirement for completeness and independence in the set.

References: Henry L. Langhaar, *Dimensional Analysis and Theory of Models.* New York: Wiley, 1951.

P. W. Bridgman, *Dimensional Analysis.* New Haven: Yale University Press, 1963.

14. *Anthropology*

Anthropologists are often classified into two types: fieldworkers like Malinowski and Mead, and "armchair" anthropologists like Fraser and Lévi-Strauss—who rely on the reports of others for their theories. Discuss the proposition that "armchair" anthropologists are studying not cultures, but anthropologists, that is, not fish, but nets.

References: James G. Fraser, *The New Golden Bough* (abridged edition). New York: Macmillan, 1958.

Claude Lévi-Strauss, *Structural Anthropology.* Translated from the French by C. Jacobson. New York: Basic Books, 1963.

15. *General Relativity*

The essential philosophical foundation of Einstein's general theory of relativity is called by physicists "the principle of equivalence" (not to be confused with any of our general systems principles). Roughly speaking, the principle of equivalence states that an observer cannot tell by any measurements whether his laboratory is falling freely in a gravitational field or is being accelerated in some gravity-free region of space. Explore the ways Einstein used this principle as a heuristic device, and relate these heuristics to techniques of systems thinking using the Principle of Indeterminability.

Reference: Albert Einstein, *The Meaning of Relativity,* third ed. Princeton, N.J.: Princeton University Press, 1950.

16. *Arboriculture: Synchrony versus Diachrony*

If we pick apples at any given time in an orchard, count their seeds, and weigh them, we get a plot of average weight versus number of seeds as shown in Figure 6.20. How can we determine whether this plot results from

a. predetermined numbers of seeds yielding fruits proportional to seed number

b. numbers of seeds increasing as fruit enlarges due to favorable environmental circumstances?

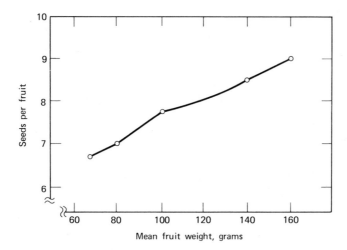

Figure 6.20. Weight of apples versus number of seeds.

Reference: A. C. Leopold, *Plant Growth and Development.* New York: McGraw-Hill, 1964.

17. *Classical Archaeology and Modern Tourism*

Figure 6.21 shows marble samples from ancient Greek quarries and certain monuments, according to their position in a two-dimensional state space based on the relative amounts of carbon-13 and oxygen-18 isotopes. Explain how an archaeologist might use this information and three mistakes that might arise in such usage. What difficulties might arise from such modern practices such as those of the Athenians, who "scatter fragments of Pentelic marble around the Parthenon each winter, in order to provide material for the insatiable pillage by tourists?"

Reference: Harmon Craig and Valerie Craig, "Greek Marbles: Determination of Provenance by Isotopic Analysis." *Science,* **176,** 401 (28 April 1972).

18. *Applied Anthropology*

Comment, in terms of the material of this chapter, on the following story:

Natives on the Pacific island of Nauru traditionally drank a strong homebrew made from fermented palm leaves. But after World War One, Nauru was mandated to Australia and prohibition was imposed. Infant mortality rose to the 50 percent level within six months. The reason? The people's natural diet was so low in vitamin B_1 that infants being nursed got the required amount of

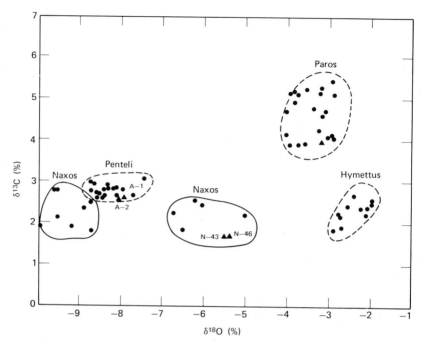

Figure 6.21. Carbon-13 and oxygen-18 variations in marble samples from ancient Greek quarries relative to the PDB isotopic standard. Triangles denote some of the archeological samples.

it only when the mother was drunk. When the natives were allowed to drink again, infant mortality fell at once to seven percent.
   Reference: Playboy, **19,** No. 6, 186 (1972).

READINGS

RECOMMENDED

1. Hans Elias, "Three-Dimensional Structure Identified from Single Sections." *Science,* **174,** 993 (3 December 1971).
2. Herbert A. Simon, "Understanding the Natural and the Artificial Worlds." *The Sciences of the Artificial.* Cambridge, Mass.: MIT Press, 1969.

SUGGESTED

1. P. W. Bridgman, *The Way Things Are.* Cambridge, Mass.: Harvard University Press, 1959.
2. Edwin A. Abbott, *Flatland. A Romance in Many Dimensions.* New York: B&N Press, 1963.
3. John M. Dutton and William H. Starbuck, Eds., *Computer Simulation of Human Behavior.* New York: Wiley, 1971.

NOTATIONAL EXERCISES

1. Show how a precise algorithm might be constructed to calculate the point of view of Figure 6.13.

2. How would the algorithm of Exercise 1 be modified to give a view in which all even digits were summed and odd digits were summed—to display the "eventropy law"?

3. Suppose we had a candy machine that took only quarters as input but allowed the selection of 10-, 15-, and 25-cent candy bars. Show how the logic of making change for this machine could be represented by an algorithm.

ANSWERS TO NOTATIONAL EXERCISES

1. The program might look like this, with v a set of 10 numbers (v_0, $v_1, \ldots, v_9$) used for calculating the ten sums at any instant in time:

  1. Repeat lines 2–7 indefinitely, for $t = 1, 2, 3, \ldots$.
    2. Calculate the next state.
    3. Set $v = (0, 0, 0, 0, 0, 0, 0, 0, 0, 0)$
     4. Repeat lines 5–6, for $i = 1$ to 100.
     5. $k = d_i$
     6. $v_k = v_k + 1$
     7. Plot $(v_0, v_1, t), \ldots (v_8, v_9, t)$.

Notice how we can write line 2 just assuming that we give the exact algorithm for state changing somewhere else—an application of the idea of decomposition, which in computing is called "making subroutines."

Notice also that we plot the sums in pairs with t, so that only five separate graphs are displayed, as shown in the figure.

2. In the above program, we would change lines 3–7 to read:

> 3. Set $v = (0, 0)$
>> 4. Repeat lines 5–6 for $i = 1$ to 100.
>> 5. $k = d_i$
>> 6. If k is even, then $v_0 = v_0 + 1$
>>> otherwise $v_1 = v_1 + 1$
>
> 7. Plot (v_0, t), (v_1, t).

3. The program might look like this:

> 1. Receive quarter and selection s
> 2. If s is empty, then return quarter and terminate.
> 3. If s is 10-cent bar, return dime and nickel.
>> a. If there is no dime and nickel, turn on NO CHANGE, return quarter, and terminate. Otherwise, give candy bar.
> 4. If s is a 15-cent bar, return dime
>> a. If there is no dime, return two nickels.
>> b. If there are not two nickels, turn on NO CHANGE, return quarter, and terminate. Otherwise, give candy bar.
> 5. If s is a 25-cent bar, give candy bar.

7

Some Systems Questions

Nothing is permanent except change.   Heraclitus

The Systems Triumvirate

But the real shift here is from a focus on organization to a focus on action, from being to behaving, from form to function, from pattern to process, from the timeless to the temporal. "Being" is the cross section of an entity in time, and those aspects of the organization which appear relatively unchanged in a series of such instants constitute the essential structure of the entity or organism. Invariance in time helps to identify the significant units of a mature system. Conversely, along a longitudinal section in time appear the transient and reversible changes, often repetitive, that constitute "behaving" or functioning, and the enduring and irreversible changes, often progressive, that constitute "becoming" or developing. And with this shift in time there occurs a shift in the entity of concern—from an object, a pattern of matter in space, to a behavior, a pattern of events in time.

R. W. Gerard[1]

We have now reached the end of Part I of our planned excursion into systems thinking. This chapter forms the interface between Part I and the rest of our journey—looking back at where we have been, and ahead to where we must go in future books.

Where have we been? In Gerard's suggestive terms of "being, behaving, becoming," we have discussed the ways we picture *being*: the notions of set, diagrams of structure, properties, boundaries, and the white box; and *behaving*: state spaces, chronological graphs, input, randomness, and the black box. We have also studied the relationship between being and behaving—how particular behavior leads to the inference of particular structure through the extraction of "properties," and how particular structure leads to the production of particular behavior through the execution of "programs."

But we especially studied all these things from the point of view of a fourth category—"believing." How, we asked, was the observer, the believer, involved in these observations? The answer came in many forms—the Eye–Brain Law, the Generalized Laws of Thermodynamics, the Generalized Law of Complementarity, the Principle of Difference, the Principle of Invariance, the Strong Connection Law, the Picture Principle, the Synchronic and Diachronic Principles, the Principle of Indeterminability. The conclusion from all these answers was that we, as observers, are entangled with what we observe, entangled in ways that leave ultimately indeterminable which is being and which is believing.

All this has been excellent preparation for the difficult territory ahead, but we have not yet raised the question of "becoming." When we are born, we are given an entire world, ready made. We do not *think* of it as ready made, for we do not think of it as made at all. "The world is so full of a number of things" that we are content to see the world of objects as a box of toys, and the world of ideas as a garden of verses. As we grow older and experience change (our teddy bear loses a leg, our family moves, our dog is "put to sleep") only then do we begin to realize the impermanence of the world. Only then do we ask the question of *becoming*: "How did things get the way they are, and why can't they ever stay the same?"

It is our experience of life that arranges these questions. As we age— as nations rise and fall from greatness, as theories flower and fade, as love grows and loved ones die—we begin to understand that "nothing is permanent except change." We are no longer so puzzled by change. Instead, we are more anxious to know, "Why does anything remain the same?"

When we are older still, and have seen not just the external world fade, but even our cherished illusions, we ask "Why do I see what I see?" Our view of the world matures from being to behaving to becoming to believing—which, of course, completes the circle.

These, then, are the three great questions that govern general systems thinking, the *Systems Triumvirate*:

1. *Why do I see what I see?*
2. *Why do things stay the same?*
3. *Why do things change?*

All general systems thinking starts with one of the three and pursues it until forced to move to another. We can never hope to find the end; we do not intend to try. Our goal is to improve our thinking, not to solve the riddle of the Sphynx. We may start where we wish, and we have chosen to start with Question 1. In a future volume we hope to explore Question 2 in detail, which will then lead to Question 3, which will certainly fill another volume. Then, if we have the stamina and the courage, we may return once again to Question 1.

But, since the questions do form a cycle, we would like to take them once around together, just so we can get a glimpse of where we are going before we get there—since we shall never really get there at all.

Stability

The general idea is always an abstraction and, for that very reason, is some sort of negation of real life. . . . Human thought, and, in consequence of this, science, can grasp and name only the general significance of real facts, their relations, their laws—in short, that which is permanent in their continual transformations—but never their material, individual side, palpitating, so to speak, with reality and life, and therefore fugitive and intangible. Science comprehends the thought of the reality, not reality itself; the thought of life, not life. That is its limit, its only really insuperable limit, because it is founded on the very nature of thought, which is the only organ of science.

Bakunin[2]

To some scientists, these words from a nineteenth-century anarchist are as the red flag to the bull, but Bakunin was no enemy of science. In spite of our present distorted view of anarchist philosophy, and in spite of Bakunin's recognition of the limitations of science, Bakunin was a fervent advocate of a truly scientific society. His words are not a condemnation, but an analysis—an analysis surprisingly modern in its thought.

We have been discussing behavior of systems, their "continual transformations." Now we must turn to the question of "that which is permanent in them," for such is the material of science today as a century ago. We must begin by stripping from the word "stable" some of its everyday connotations that would only serve to mislead our general systems thinking.

In the first place, of course, "unstable" is one of those behavior words that are justifiably applied to even a single instance of undesirable behavior. Throwing one bomb makes a man an anarchist; falling down once makes a building unstable. If a building falls down it is—or was—unstable; but everyday speech errs in assuming that buildings can do but two things, stand still or fall down.

We often confuse the word "stable" with the word "immobile." Most of us would agree that the Empire State Building is stable, and for that reason we are surprised to find that the top sways from side to side on moderately windy days. Stable does not mean complete lack of change, any more than anarchy means complete change from order to chaos. Stability, rather, means change *within certain limits*—which is, really, neither more nor less than anarchy means.

In a sufficiently strong wind, any building can be blown down. Must we conclude that no building is stable? Not at all. Stability not only implies limits to the changes in the system, but also implies limits to the disturbances that the system is supposed to withstand. Thus, when we speak of stability, we are speaking of two things: a set of *acceptable behaviors* of the system and a set of *expected behaviors* of the environment. We are, to put it another way, defining a region of the environment's state space and a corresponding region of the system's. For example, we might define stability of a tall building as "swaying not more than 10 feet from the perpendicular in winds of up to 90 miles per hour."

Stability, in our definition, is a relationship between system and environment. In a closed system we could imagine such a thing as absolute stability, but that only pushes the question out to the isolating boundary. How are we to construct the absolutely stable boundary needed to close a system absolutely? The physicist realizes this limitation, so his concept of stability relates to something called "small disturbances." The system is opened *a tiny bit*, and its behavior is then observed. If the effect of the disturbance disappears after a while, the system is said to be stable, whereas if it is magnified, the system is said to be unstable. The classical example is the pyramid. When standing on its base, the pyramid returns from a small nudge to its original, stable position. If standing on its apex, however, the pyramid will be sent tumbling by any small disturbance from its original, unstable, position.

There is something unsatisfyingly circular in the physicist's definition of stability, for how does he know what is a "small" disturbance except in terms of his test for stability? The absolutism has merely been transferred from the system to the environment, which now contains absolutely "small" disturbances. Although this kind of argument may suffice for scientists working with approximately closed systems, it merely misleads those who have to contend with openness that they cannot define away in the laboratory. In particular, it may misdirect attention and cause us to seek stability "in" the system, rather than as a relationship between system and environment.

Such absolutist thinking about stability is particularly dangerous in ecology, and accounts for much of the ravaged condition of our land-scape. For example, a tree growing in a dense forest stand derives part of its ability to withstand wind from the surrounding forest. When log-gers think of stability as being "in the tree" and cut down many trees in hopes of leaving a few stable trees standing, the isolated trees are often blown down in the first substantial storm[3].

This sort of absolutist thinking is so common that people ought to ap-ply the Axiom of Experience and begin to judge the future by the past. But if the past is any guide, they won't. One need only open one of the 42 volumes of the American Environmental Studies Collection[4] at random to find similar moral fables—like that of the passenger pigeon, which once existed in the *billions* in North America, but which exists no longer. We read in the "Report of a select committee of the Senate of Ohio, in 1857, on a bill proposed to protect the passenger pigeon":

> The passenger pigeon needs no protection. Wonderfully prolific, having the vast forests of the north as its breeding grounds, travelling hundreds of miles in search of food, it is here today and elsewhere tomorrow, and no ordinary destruction can lessen them, or be missed from the myriads that are yearly produced.

And what, pray tell, is "ordinary destruction," if not a synonym of "small disturbance"? When it comes to the preservation of irre-placeable resources, it will simply not do to wait until we actually suc-ceed in destroying the system before defining what we mean by "small disturbance."

But absolutism is useful, and thus dies hard. A more sophisticated attempt to preserve some notion of stability "in the system" is the con-cept of a "linear system." A system is linear with respect to a particular input if increasing that input merely changes something in the system by the same amount. In functional form, if the "response" of the system is given by

$$R = f(I)$$

where I is the input, then if f represents a linear system,

$$f(2 \times I) = 2 \times f(I)$$
$$f(1000 \times I) = 1000 \times f(I)$$

or, in general:

$$f(a \times I) = a \times f(I)$$

for any conceivable value of a.

The linear-system concept, like the closed-system concept, is an extremely valuable approximation, one that should ultimately be studied with care by all systems thinkers[5]. Its importance for stability lies in its removal of the relative concept "small disturbance," because as long as the disturbance is *finite*, the behavior will be of the same *sort* as before, even though bigger. For instance, if my hi-fi is linear and I turn up the sound, the music is louder, but not distorted—and I can keep turning up the sound without ever hearing anything that was not in the original music.

Unfortunately, the linear-system concept, though helpful in systems thinking, simply pushes the absolutism further into the shadows. No system we know is strictly linear. If I turn up the volume far enough, distortion of the music is bound to result. My hi-fi may be designed to prevent me from turning the volume knob that far, or may simply stop working to prevent me from damaging the loudspeaker, but these are both nonlinearities intended to keep the music in the linear range. What the linear-systems approximation really says, then, is that the system is linear "as long as it stays within reasonable bounds." But what are these "reasonable bounds," and how can I discover them without first reaching nonlinearity?

Another problem with the linear-system approximation is that a system *need not be linear to be stable.* By confining our attention to linear systems we would fail to notice large classes of systems that display stable behavior. We use the linear-system model not because we find the world to be particularly linear. As with other approximations, we easily fall prey to believing our model more than the empirical world, with the result that we may be unable to "see" nonlinear systems, just as we are unable to "see" nonthing systems.

For example, linear systems have the convenient property of "superposition." If we put two of them together, the result is a linear system, at least if we stick to the rules of "putting together." For example, if we have two systems:

$$R = f(I) \qquad S = g(I)$$

both of which are linear, then we can put together

$$T = R + S = f(I) + g(I)$$

which is also linear. Or, going backwards, we can decompose a linear system into systems that retain the property of being linear.

Superposition is a handy property, as far as it goes, but it may lead the careless into fallacious arguments about nonlinear systems. For example, W. J. Cunningham,[6] in an otherwise sophisticated article about what engineers mean by stability, makes the following argument:

Many large engineering problems can be broken into smaller portions. While the ultimate test of an airplane is whether it functions and flies as a whole, there are many smaller systems within the airplane which can be considered separately. . . . If the airplane is to operate properly as an entire system, each of the small systems must operate properly. The entire system can hardly be called stable unless the component systems are themselves stable.

But, of course, what Cunningham is doing is arguing by unconscious analogy with linear systems, and making a Decomposition Fallacy. When we admit nonlinear systems into our universe, we can easily put together stable systems out of parts that would be unstable outside of the whole. The tree in the forest was one, and the individual passenger pigeon was another. While we humans prefer to do our mating in private, passenger pigeons evidently would not mate unless they were in a tree with thousands of their fellows. Thus, the "yearly production" of passenger pigeons was not a linear function of the number of pigeons at all. Half as many pigeons did not produce half as many new pigeons, and after halving enough times, they eventually produced none at all.

Another conceptual difficulty we wish to avoid is the idea that "stability" is somehow equivalent to "goodness." Consider a coal mine that caught fire fifty years ago and has been burning ever since—displaying high stability, but not much "goodness." Moreover, since stability and goodness are both defined relative to a particular point of view, it is easy to see how the same situation could at once be "stable and good," "stable and bad," "unstable and good," as well as "unstable and bad." The form of government may remain stable, even though various officeholders shuffle in and out. From the conservative point of view, the government is stable and good—while the radicals think it is stable and bad. The deposed officeholders think it is unstable and bad, while the poor abused citizens think it is unstable and thereby good—since the rascals get thrown out eventually.

Still, we do tend to feel that stability and goodness are related, and they are—in our minds. How does this association come about, and why is it so universal? When we seek explanations for such admittedly universal impressions, we should look for some equally universal experiences from which the impressions arise. We are more likely to *notice* changes than things that are unchanged. Furthermore, among the things we do notice, the ones that cause us pain or discomfort are more likely to stand out as individual impressions. Thus, when a change takes place, the ways in which we feel we are worse off generally make a stronger impression, so we begin to equate change with badness and, by implication, stability with goodness.

Perhaps each of us has an ideal world in mind in which only the "bad" things change while all the "good" things stay the same. But our world is not built that way, if only because we change our definition of what is good as time goes by. In the same way, we can change our definition of what is stable, and a system that was once "stable" can become "unstable" simply by our changing ideas as to what should be the range of its behaviors or what should be the range of environments. The change may come about gradually, as in the case of a parent who comes to accept new behaviors of his children; or it may be rapid, probably in response to an actual event, like extinction of the passenger pigeon.

Suppose, for example, that a building were to fall down on a day when the wind velocity reached 110 miles per hour. The owners of the building and the people injured in the event would accuse the architect of having designed an "unstable" building. At such a time, it will not do much good to remind the owners that they had approved plans in which "stability" was defined in terms of maximum winds of 90 miles per hour. The occurrence of a 110 mile-per-hour wind has changed their conception of stability, even though before the fact they might never have approved of the extra expense to build for a wind velocity that had not been previously known.

In the final analysis, all three parts of the concept of stability—system, environment, and critical limits—depend on the observer. But then how do we account for the feeling that we have that stability is somehow central to our thinking about systems? The answer was nicely put by Parsons and Shils[7] when they said:

> ... if a system is to be permanent enough to be worthy of study, there must be a tendency to maintenance of order except under unusual circumstances.

In other words, an arbitrary selection of variables out of the world does not have to exhibit stability, but the less stable it is, the less chance it has of being "worthy of study."

Perhaps it would have been better to say "capable of being studied," rather than "worthy of study," for scientists will go to unbelievable ends to make or find the needed stability in something they *want* to study. Sometimes they shrink the time scale—an isotope that is "stable" for a millionth of a second is quite adequate a system for the physicist. Figure 7.1 is a photograph of a *pulse of light* (!) photographed as it passed through a vial of water using a shutter speed of 20 *trillionths* of a second to make it "permanent enough to be worthy of study"—using an experimental arrangement such as we proposed for studying cars coming out of toll booths.[8]

By these and other devices, scientists have been very successful at abstracting or creating points of view that elevate the importance of the more constant parts of the system—so successful that they forget that this constancy is largely a matter of choice, not chance. Darwin[9] recognized this situation over a century ago when he observed that:

Authors sometimes argue in a circle when they state that important organs never vary; for these same authors practically rank that character as important (as some few naturalists have honestly confessed) which does not vary; and, under this point of view, no instance of an important part varying will ever be found; but under any other point of view many instances assuredly can be given.

Figure 7.1. Light photographed in flight.

Darwin was, of course, thinking of anatomists and naturalists, but it is interesting to observe the same kind of circular argument "honestly confessed" over 100 years later in the field of anthropology by such a noted theorist as Julian Steward[10] (author's italics):

> The present statement of scientific purpose and methodology rests on a conception of culture that needs clarification. *If the more important institutions of culture can be isolated from their unique setting so as to be typed, classified, and related to recurring antecedents or functional correlates, it follows that it is possible to consider the institutions in question as the basic or constant ones, whereas the features that lend uniqueness are the secondary or variable ones.*"

What Steward is expressing is a heuristic principle, our Principle of Invariance, but it may sound like a law of nature if we do not read him carefully. He states: "it is possible to consider . . . as basic or constant," from which we might infer that "basic is constant," or "constant is basic." What he really must be saying is "constant is studiable"—the Principle of Invariance—all of which is simply preparation for one of our triumvirate questions:

"Why do things stay the same?"

Survival

If you look at automata which have been built by men or which exist in nature you will very frequently notice that their structure is controlled to a much larger extent by the manner in which they might fail and by the (more or less effective) precautionary measures which have been taken against their failure. And to say that they are precautions against failure is to overstate the case, to use an optimistic terminology which is completely alien to the subject. Rather than precautions against failure, they are arrangements by which it is attempted to achieve a state where at least a majority of all failures will not be lethal. There can be no question of eliminating failures or of completely paralyzing the effects of failures. All we can try to do is to arrange an automaton so that in the vast majority of failures it can continue to operate. These arrangements give palliatives of failures, not cures. Most of the arrangements of artificial and natural automata and the principles involved therein are of this sort.

John von Neumann[11]

Why does a system survive? From the long-range viewpoint, a system survives because systems that do not survive are not around to be thought about. The systems we are accustomed to seeing are systems that have been *selected* from all systems of the past; they are the best "survivors."

We should quickly observe that surviving is a truly remarkable thing for a system to do. Our view is biased because we most often see systems that are good survivors, but the vast majority of systems do not survive for long—over any time span we wish to choose. At the level of biological individuals, we have reason to believe that nobody lives forever. The oldest living things we know, the bristle cone pines, are about 4000 years old. If we choose *populations* as our systems, so that the systems survive even when individual members die, the situation is not much better. Since life began on this planet, over 90% of all species that ever lived are now extinct—and there are but few species like the cockroach, which has been around about 300,000,000 years. Human organizations are even more puny. Most new businesses fail within five years, and it is hard even to think of a business which has been around for more than a few hundred. Organizations like the Roman Catholic Church, which is not quite two thousand years old, are of the greatest rarity.

Survival, then, is far from being a trivial property of system behavior. It is a property that every system must have for us to study it, and a property that not every arbitrary collection is likely to have. Consequently, it is important that we have a pellucid understanding of what we mean by survival. Since survival is the continued existence of

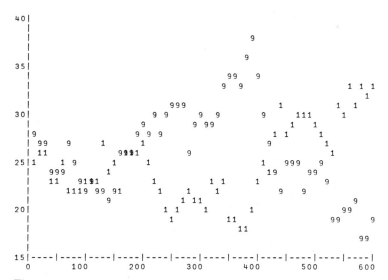

Figure 7.2. Behavior in region *S*1.

a system, we must, if we are to be precise about survival, examine the meanings of "continued" and "existence."

"Continued" refers to the length of time a system has to exist in order to be worthy of study. How long this must be is a question of relative time scale between system and observer, and thus relates, at least indirectly, to the typical length of time the observer survives. In the case of man as an observer, the effect of time scale is not difficult to ascertain. We do not, for instance, ordinarily think of plants as moving about under their own power, but if we watch a plant through the quickened time scale of a time-lapse motion picture, we see it as writhing about in a most agitated manner. Through the device of slow-motion photography, we can begin to empathize with worlds such as the microbiological where things otherwise are born and dead before we can apprehend them.

Our white-box system can be used to illustrate time scale. Recall the four types of OCCULT clubs, *SX, SI, SV,* and *SVI,* so named after the equifinal state to which they led under random input. While we portrayed the approach to states *SX, SV,* and *SVI,* we only *talked about* the inevitability of reaching *SI.* If we do show the behavior from such a state as (0, 25, 0, 25, 0, 0, 0, 25, 0, 25), we see something like Figure 7.2, which does not at all seem to be approaching *SI* or anything else.

The reason the behavior does not seem to approach *SI* is plain to see in Figure 7.3, where we have started the club in a state (0, 99, 0, 0, 0, 0, 0, 0, 0, 1), which is the state that must precede *SI.* A very special input such as the fifteenth line of Figure 6.4 [(38, 38), where $d_{38} = IX$] would

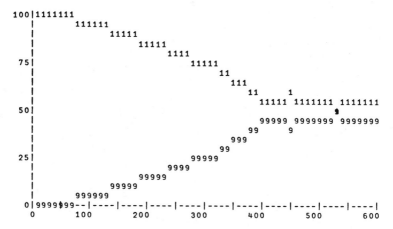

Figure 7.3.   Coming close to *S*1, but not quite making it.

be required to move the system to *SI*—a member of caste IX must "teach himself" to become an I. Since that input was not sent from Headquarters, the club moved farther and farther away from *SI*, though remaining in the subspace of all I and IX. In other words, there is very little chance of getting into *SI*, even when the club is "close"— and to get that close would have already required a very long sequence of improbable inputs. If we had not *started* the club close to *SI*, by choosing only the most advanced as members, we would not likely have seen it get so near, let alone actually get into *SI* on its own.

How long it takes to reach this state of perfect oneness depends on how the Grand Mystagogue controls the inputs. If his inputs are, in fact, "essentially random," it will probably be a very, very long time before we see a club actually get to *SI*—though once there it shall surely stay.

How long is very, very long? If our computer can make a thousand state transitions a second, we should probably have to wait much longer than the age of the universe to see *SI*. So, though we know that the system must, by force of logical necessity, reach *SI*, we would never characterize it in that way if we knew it only through observation of its behavior. In that case, Figure 7.2 would be a "typical" picture of the behavior, and we would say that this pattern of behavior is stable, even though it is doomed to *eventual* extinction. And if, after a long time, we chanced to observe a system reaching *SI*, we would undoubtedly say that the system we knew had failed to survive.

Identity

The most fundamental concept in cybernetics is that of "difference," either that two things are recognizably different or that one thing has changed with time.

W. Ross Ashby[12]

To exist is to have an identity. Identity is indeed synonymous with viability, for nothing remains to be identified that is not viable, and a thing that changes its identity passes out of existence. But to have an identity is to have an identifier, and hence comes the difficulties of saying when a system exists or ceases to exist. Did Rome fall in 476? Some historians say so, but several million Romans live in the city today. Did Dr. Jekyll cease to exist when Mr. Hyde was roaming the streets? It all boils down to a question of how we establish the identity of a system.

Looking out my kitchen window on an early spring day, I saw a blue jay in the cherry tree. I recognized it as a blue jay because it was blue. I am not a professional (or even an amateur) bird watcher; if someone showed me a red blue jay, I would call it a cardinal. Furthermore, I feel safe in saying that most people would make the same identification. But an ornithologist sees the matter differently: for him color is not the major identifying property, or variable, of blue jays. No doubt he would be surprised and delighted to find a red blue jay, but he would never mistake it for a cardinal; to him it would be a sport, a mutation, but still a blue jay. Now, who is right, the ornithologist or I? We are both right, of course. We may argue about whose set of identifying variables is a better one, but we are likely to come away from the argument holding our original views. My criterion is perfectly adequate for my poetic purposes and perfectly inadequate to the ornithologist's prosaic ones.

Perhaps this example seems ludicrous. It was chosen for clarity, not for subtlety, so as to avoid prejudicing the argument, as might have been done with numerous classical examples. When did the first man appear on earth? What happened to the Standard Oil Company? Who is a Negro? Does the revolutionary government have to honor agreements made by its predecessor? Is space empty or filled with an ether? In all these cases, the argument has meaning only because there is no agreement about the identifying variables of the system in question. Still, we are all confident that we know a man when we see one, know what constitutes a company, and can recognize Negroes, governments, and empty space. Only when our fuzzily defined concepts are challenged by actual cases do we see how fuzzy they usually are; for in everyday life, we have no need for more careful delineation.

But surely there are some systems on which all observers agree? For example, the atoms of a single element are surely identical to all observers—aren't they? Most emphatically not! The physicist may pretend that they are for various purposes, but for other purposes he uses tools such as Mössbauer spectroscopy to study how the energy state of the nucleus is modified by the surrounding atoms in a crystal. This change of definition of "same" plagues all discussions of change, from physics to philosophy, because in fact we carry not just one definition of "same" but a large *set* of definitions.

We can clarify our thinking about "sameness" if we follow the path of those who try to program computers to recognize sameness and difference—a subject called "pattern recognition"[13], or, in the more specific case of visual images, "picture processing"[14]. When presented with two pictures that are discriminably different, the computer can answer the question of "sameness" by processing the two pictures in such a way as to reduce them to a standard—or "canonical" or "normal"—form.

For instance, in Figure 7.4, we see several "letters." Are they the same letter? If we answer "yes" we have transformed the letters mentally into a canonical form, as shown in Figure 7.5. Such a transformation is another application of the Invariance Principle. In this case, "sameness" of letters is a property that is not affected by change of size or rotation on the page.

But, of course, it is not that easy. Are the letters in Figure 7.6 "the same"? Rotating one gives the other, but one is an "N," the second is a "Z," and the third is—well, who knows what the third is? Sometimes it is an N, sometimes it is a Z, sometimes it is neither, and sometimes it is either (Figure 7.7). Each set of rules for normalization gives us a different definition of "same," which we can apply with equal facility. The deep problem is not whether two things are the same, but knowing if everybody is talking about the same kind of "same."

"Difference is the most fundamental concept in cybernetics"—and in general systems thinking as well. We must never forget that it is also the most *difficult* concept, a caution to be remembered throughout the following discussion. We shall assume, for this discussion, that we have agreed on some definition of "same" and are going to try to *identify* a system.

For example, we may agree to define sameness for our OCCULT system as indicated in Figure 6.17, so that there are four kinds of systems: *SX, SI, SV,* and *SVI.* The reduction to one of the four canonical forms could be accomplished by the following program:

1. s = count of ones, threes, sevens, and nines.
2. If s = 100 then this is *SI.*
3. Otherwise t = count of fives.

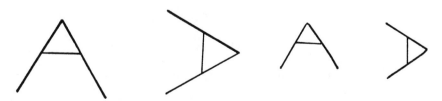

Figure 7.4. Are they the "same" letters?

ORIGINAL   CANONICAL FORM

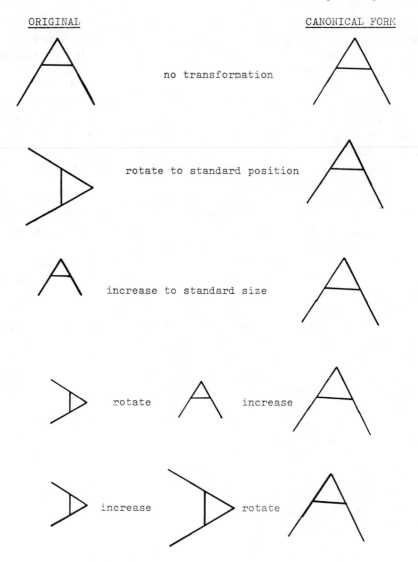

Figure 7.5. Transforming to canonical form.

4. If $s + t = 100$ then this is *SV*.
5. Otherwise u = count of zeros.
6. If $u = 0$ and $t = 0$ then this is *SVI*.
7. Otherwise this is *SX*.

In other words, identification here reduces to a single observation of

Figure 7.6. Are they the "same" letter?

the state followed by transformations to determine the region of the state space in which that state is found.

Actually, we can only identify with certainty in this fashion in a unique case—the case of a closed, state-determined system for which we have a complete view. Because of our involvement with this special metaphor, we have a tendency to imagine that identity is always determined by just such an instantaneous observation, but our customary procedure is to observe behavior over some period of time. How long a period of time depends on the consequences of making a mistake—a long engagement may be prudent where divorce is prohibited.

Figure 7.7. Is it N or Z?

But suppose we use this simple view of the process of identification. If we are actually observing a closed, state-determined system, the identification cannot fail. If the system has input, however, this simple method will give correct identification only if the state happens to stay within the range we have set, as it would with our simple system. In general, however, because

$$S_{t+1} = F(S_t, I_t)$$

there is the possibility that some I will come along and send the system out of the identity region. To return to our analogy, some clumsy house painter may splash red paint on the blue jay.

What are the chances of such an event occurring? We cannot, of course, calculate the chances of such a bizzare event, but we can see what they depend on. If we conceptualize the functional relationship as in Figure 7.8, we see that the survival of the system depends on the input I from the environment, but also on F, the way the input is interpreted or transformed by the system. The system survives if it is built in such a way that all the identifying properties are stable.

By "built in such a way," we mean partly the "laws of nature"—the rules by which *any* system is built—and partly the special system laws—the structure of that *particular* system. Thus, if part of the "program" of the blue jay is

1374. If someone splashes red paint on you, fly to a pail of turpentine and take a bath.

then the bluejay will survive, in my perception. Or, its program might have contained

3502. If someone splashes red paint on you, stay out of this particular cherry tree until the color wears off.

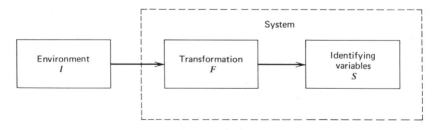

Figure 7.8. The problem of maintaining identity.

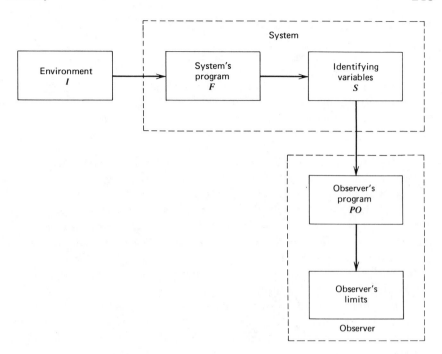

Figure 7.9. The general problem of maintaining identity.

to the same effect. Seen in this way, the system survives if it happens to have a suitable transformation for the environment in which it finds itself, or, rather, in which the observer finds it.

To be more precise, we have to portray the general problem of maintaining identity as shown in Figure 7.9. There, the observer is put into the picture along with *his* program for determining whether or not the system has maintained its identity. In this more general view, the question of survival depends on

1. What the environment does
2. How the system's program transforms the environment
3. What variables are involved in the identity
4. How the observer's program operates on those variables

Regulation and Adaptation

> Oh chestnut tree, great rooted blossomer,
> Are you the leaf, the blossom or the bole?
> O body swayed to music, O brightening glance,
> How can we know the dancer from the dance?

William Butler Yeats[15]

Where do systems come from? In the case of the painted blue jay, failure of one system to survive leads to the creation of a new one: one less bluejay equals one more cardinal. But in other cases, a system that fails to survive may not be replaced: it just fails to survive. As time goes by, will there be fewer and fewer systems, as unsuccessful transformations fall by the wayside—both in systems and in the observer's program? Not if there is a source of new transformations. In that case, we need not see a world growing progressively less and less interesting as all our old familiar systems die off.

Where does the transformation come from? Where is it in the system? In general, we cannot say, but in our white box, we know exactly where the transformation is found. The transformation is the program.

Are different programs possible? Of course, but as far as we yet know, only if we intervene to write them. Well, not precisely, for recall that we could allow a *parameter, n,* to be changed by the input, thus creating a different simulation. We can carry this simple idea much further.

In our white box, we had the instructions

> 0. Repeat lines 1–4 indefinitely.
>   1. Get next pair (i, j).
>   2. $t = d_i \times d_j$
>   3. d_j = last digit of t.
>   4. Display $(d_1, \ldots, d_{100})$.

There are several types of instructions here—"repeat," "get," "assign a new value," "display output"—but there are still other operations a computer can handle. In particular, we can choose an arbitrary input pair (73, 15) and then insert an instruction that says

> 1.5. If (i, j) is (73, 15) then change line 2 to read as follows:
>   2. $t = d_i \times d_j + 1$

In other words, the program can change *itself*, just as an OCCULT club might revolt against the Cabalist hierarchy and change its own rules.

Inside the computer, the program is coded into numbers and stored in precisely the same way as either "state" or "environment" is stored. All three boxes of Figure 7.8 are, in our simulation, represented as numbers stored in the computer's memory, and there is no reason to suppose that one of them is more permanent than the others. In particular, the program itself can be variable, so that instead of regarding the transformation as some mysterious "unknown function F" in

$$S_{t+1} = F(S_t, I_t)$$

we can decompose S into (program, other variables), or (P, V). All that remains of the mysterious F, then, is the set of rules, built into the computer, by which the program is to be interpreted. These are the only immutable part of the simulation, but as Simon said, "almost no interesting statement that one can make about an operating computer bears any particular relation to the specific nature of the hardware."

For the inner world of the computer, then, the hardware represents the "laws of nature," the stage on which the play of simulation is carried out. Though the simulation "depends on" this hardware, the play's the thing, not the stage management. Therein lies one beauty of the computer as a stage for simulation—it allows us to build any kind of set, any kind of fantasy world we might like to study by the method of the white box.

In function form, what we have been saying is

$$S_{t+1} = H(S_t, I_t)$$

where

$$S = (P, V)$$

and where by H we mean the "hardware" or "immutable laws of nature," thereby drawing attention to the fact that everything else *might* change, including the program. Moreover, everything *universal* is expressed in H, while everything *specific* to this system is stored in S and I. Therefore, the source of any "special systems laws" will have to be found in *them*, and not in the hardware.

Up to now, when we have spoken of the system's transformation, we have implicitly assumed it to be fixed. Now that we have succeeded in placing the transformation on the same level as the other aspects of the system, we are ready to apply the Principle of Indifference. To understand change, we shall have to consider the possibility of *change in the transformation itself.*

Changes in transformation are no strangers to us. They may be less familiar than things that stay the same, or they may simply be harder to think about, but they are around us all the time. The most obvious example is what we call "learning"—for we can imagine that what we know how to do is stored in our brains much as programs are stored in the computer. By learning, we change our identity, for "to understand is to change, to go beyond oneself." It is not skeletal structure that makes the difference between a doctor and a lawyer. But we can also alter our physical structure and thus change our identity: our bodies can "learn" to be pole-vaulters or pole-climbers. Or, we can extend our bodies with new tools: a "peasant" trades his hoe for a tractor and becomes a "farmer."

Any system can undergo change in that part of itself that determines the way it transforms its input—changes in its program, whether that "program" takes the form of bones, buildings, or beliefs. Therefore, there is not just one way of preserving the identifying variables, but two. If the state of the system is (P, V), the program and the rest of the variables, then it can preserve its identifying variables either with a fixed P or by varying P itself.

In recognizing the possibility of changing the program, we solve the problem of consuming possible transformations. In our white box, for instance, we previously recognized 10^{100} states, using only 100 digits of the computer's memory. How many programs are there? Although our program was rather short (only five lines) programs could be of any length, up to the point where they fill the computer's memory with program text. If the memory has space for 1,000,000 digits, then it could hold a program 1,000,000 digits long—perhaps 10,000 lines. Any different combination of 1,000,000 digits would be a different program, so there would be $10^{1,000,000}$ different programs, transformations, stored at different times in the same computer. Each of these programs will have some characteristic effect on the 100 digits we have singled out for attention, but not all behaviors will differ. Thus, each program we observe is one of a *set* of equivalent programs, each of which looks the same to black-box observation over a certain period.

To illustrate this principle, consider the behavior displayed in Figure 7.10. Although this behavior seems similar to that in Figures 6.13 or 6.14 (note the new time scale), something has changed. Instead of settling down to SX and remaining in SX, the OCCULT club waits a while and then jumps out of SX into some random state. This graph

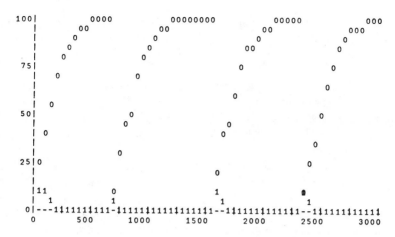

Figure 7.10.　A chronological graph of the white box.

was produced by adding the following line to the original program:

> 1.5. If state is *SX* then if $i = j$, then choose an entirely new initial state at random.

Notice that as long as the club has not "settled down" to *SX*, line 1.5 is entirely inoperative, and the club will not be reorganized. In this case, the presence of the rule cannot be detected in any way from black-box observation of the white box. *We* know it is there only because we put it there. In an OCCULT club started in *SI* or *SV* or *SVI*, line 1.5 will be *completely* inoperative, and we could never distinguish the two programs.

But there is another way in which line 1.5 can be rendered inoperative. Since the club must receive an (i, j) with $i = j$ before restarting, what happens if the input is constrained so that "self-teaching" never occurs? What happens is what we saw in Figure 6.13— a behavior exactly like the original system without line 1.5. Indeed, Figure 6.13 was "really" generated in this way, and not by our original program, for Cabalists do not believe in meditation, so Mr. O'Teric never permitted self-teaching.

By the Principle of Indeterminability, we know that we are unable to tell the difference between the club with line 1.5 but no meditation, on the one hand, and the original club not allowing meditation, on the other. Therefore, we should not be surprised when Mr. O'Teric removes the constraint against meditation so as to permit clubs to reach *SI*—

and we see a sudden jump in Figure 6.13. We *should* not be surprised, but we are. Why? Because we tend to think of the transformation as a separate, or separable part of the system, a part that, moreover, does not change.

Up until now, we have been assuming that the system was identified by the states of certain variables, and not by the display of certain behaviors. In Yeats' words, we have identified by the dancer and not the dance. Now it is time to drop that assumption. But must we really drop it? How *can* we know the dancer from the dance?

Since we have now made the program part of the state, (P, V), identification by behavior is in theory the same as requiring that the program P be one of a set of programs that produces behaviors in a certain range. In other words, the program must be stable. The system may change programs, but not enough to move the behavior out of the range by which we identify it. How much change in programs does this requirement permit? As our computer programs demonstrate, there is no *necessary* relationship between the size of a program change and the consequences it produces. The tiniest change can entirely change the behavior displayed, or great changes can have no noticeable effect.

Consider a program containing this test:

492. If $x + 1 = y$ then follow procedure A. Otherwise follow procedure B.

Suppose, in the memory of the computer, the "+" is changed to a "−," giving

492. If $x - 1 = y$ then follow procedure A. Otherwise follow procedure B.

In terms of information stored in the white box this change from plus to minus is the smallest possible change: we say it is a single "bit" of information. Yet in terms of the outside effect—what the black-box observer sees—the change could be as big as the difference between procedure A and procedure B, for any arbitrary procedures.

This example is not fatuous. Computer systems have failed on just such a single bit change. In biological organisms, the change of a single molecule in the genetic material can lead to the production of offspring entirely different from the parent. In legal matters, the insertion or deletion of a single comma can completely reverse a situation. The story is told of a man in New York who negotiated the purchase of a company in England, before the days of the transatlantic telephone. His agent cabled the offering price of $14,000,000. This was more than he was prepared to pay, so he cabled back the message

NO, PRICE TOO HIGH.

In the course of transmission, the comma was omitted, so the agent dutifully closed the deal at $14,000,000.

Such stories capture our fancy because we believe that small changes in the white box should lead to small changes in black-box behavior. We believe so because we live in a world surrounded by systems whose "structure is controlled to a much larger extent by the manner in which they might fail and by the precautionary measures which have been taken against their failure." The man was a fool to risk millions on the correct transmission of a comma and should have splurged on a few extra words. Biological organisms may be entirely changed by a single mutation, but elaborate arrangements are made to protect the germ plasm from such change and to nullify its effect if it does occur. Therefore, our experience of the world is that the *Law of Effect* usually holds[16]:

Small changes in structure usually lead to small changes in behavior.

or, put in our terms:

Small changes in the white box usually lead to small changes in the black box.

Because of our experience with the Law of Effect, we tend to partition systems into a fixed and a variable part, in which the fixed part—or "structure"—is the "source" of its behavior. Steward's statement, already quoted, is a rather typical expression of this stance, which we find almost everywhere in scientific writing. Take, for example, this passage from Herbert Spencer[17]:

It is the permanence of the relations among component parts which constitutes the individuality of a whole as distinguished from the individualities of its parts.

In other words, in this view we partition the system into two sets of variables, P and V. In V we see the "behavior," or "function," which is the variable functioning of the permanent "structure" P. Although we may identify the system by its functioning, this is only a convenience, for the "real" identity lies in "the permanence of the relations among component parts"—the "structure."

Complementing this "structural" view is the "behavioral" view, which says that the only way we know "structure" in the first place is by observing behavior. For the promoters of the behavioral view, the Law of Effect can be stated in reverse:

Small changes in behavior will usually be found to result from small changes in structure.

We might call these two stances "white box" and "black box," but they appear under many names in the history of scientific controversy. In biology, we have the anatomists, who try to understand change through statics—behavior through being. On the other hand, we have the ethologists, who try to discover what is constant through what changes—being through behavior. At another level in biology we have the molecular biologists versus the systematists; in physics we have mechanics versus thermodynamics; in psychology, the physiologists versus the behaviorists; in art, the linear versus the painterly. When will scientists finally discover that there are complementary ways of looking at the world?

The concepts of "regulation" and "adaptation" come to us from both sides of the white–black–box controversy, and therefore depend for their clarity on the clarity of partition between P and V. In the white-box view, the system is regulating if through a fixed P it keeps a stable V; in the black-box view, a stable V is evidence that P is more or less fixed. Though regulation does not imply that the system is unchanging, it does imply that the changes are "frequent," "small" or "unimportant," "reversible" or "cyclic," and in the "variable part" or "functioning" of the system.

In the white-box view, adaptation involves a change to P; in the black-box view, adaptation is seen through "significant" changes in behavior. These changes, in either view, are almost by definition "infrequent," "large" or "important," "irreversible" or "progressive," and in the "fixed part" or "structure" of the system. But in any case, the changes to P are not so large as to change the identity of the system—in that case the system would fail to survive.

All of the terms used in differentiating "regulation," "adaptation," and "failure to survive" are relative terms in diverse absolutist disguises. In point of fact, there is no absolute way of distinguishing regulation from adaptation or adaptation from failure to survive. We do not generally *partition* into (P, V), but only make a rough separation in which the same variable may be found in both "parts."

Once we have discovered the source of trouble, many of the major problems that have plagued philosphers and scientists simply disappear. In biology, does the species of fish that starts to be able to breathe air become a new species or is it simply adapting? In anthropology, does the group of people that adopts the language and some of the other ways of another group become a new culture, an adapted culture, or part of the culture from which the adoptions are taken? In organization theory, does the acquisition of new tasks make a new organization or is it an adaptation of the old organization?

Since the method of breathing is part of the transformation as well as part of the identity of a "fish"; since a language is part of the transformation and part of the identity of a "culture"; and since the tasks performed are part of the transformation and the identity of an "organization"; none of these questions can be solved—since the definitions of adaptation and preservation of identity rest on a false assumption of perfect partition.

In recent years there has been an upswelling of interest in the study of "adaptive machines"[18]—a term that antagonizes students of "natural" adaptation. If we build a simulation of an adaptive system, we must build it with a program that may vary. But if the system then shows "adaptive" behavior, critics will argue that the system is not adapting, but regulating, since the program was *designed* to be variable. Therefore, any changes exhibited are not to the original *structure* of the system, so the system is not adapting—since adapting means changing the structure.

Nevertheless, we might want to accept this way of talking as being useful in certain cases. For example, we might be led to ask whether human learning is "adaption," since the structure of the brain incorporates the possibility of change. Or what about biological adaptations, since the structure of the genetic mechanism clearly incorporates provision for change? But what is the point? "Adaptation" and "regulation" are not "real"—they are but tools of thought. Because they are powerful tools, they are often misused, but that is no excuse for discarding them or defining them away. Rather than throwing out the tools, let us try to learn to use them safely and effectively.

The Used Car Law

Stress is the state manifested by a specific syndrome which consists of all the non-specifically induced changes within a biological system.

The elements of its form . . . can express the sum of all the different adjustments that are going on in the body at any time.

Hans Selye[19]

Complementary viewpoints may engender confusion among adaptation, regulation, and loss of identity—confusions compounded by the complementary relationship that exists between regulation and adaptation within any one view. Consider, for instance, the problem facing a car owner who is a little pressed for cash. His operating expenses, which may be thought of as the cost of regulation, consist mainly of gas and oil. They are like regulatory expenses because they are incurred periodically in the "normal" operation of the car.

If he keeps a careful record of these expenses, he may notice from time to time that the gas mileage is declining. Eventually, it will be profitable to have the engine tuned because of the resulting savings on gas. He may also begin to notice oil consumption, at which point it may pay him to install piston rings, have the cylinders rebored, or maybe even buy a new engine. Now, the reason for taking these rather drastic steps, which could be considered adaptations, is that the cost of regulation has increased to the point where they are economically justified. By adapting, the cost of regulating is reduced.

Of course, there may come a time where nothing short of a change of identity—buying a new car, perhaps—will suffice to keep down operating costs in the face of the ravages of time. But before that time, the life of the car will consist of long periods of slowly declining regulatory ability, interspersed with rapid but large adaptive changes. This sort of life history is so common among systems, for reasons that will eventually be made clear, that we can elevate the underlying principle to a general systems law—the *Used Car Law*.

In general terms, rather than in terms of used cars, the law states that:

1. A system that is doing a good job of regulation need not adapt.

2. A system may adapt in order to simplify its job of regulating.

Examples of the Used Car Law are ubiquitous. The farmer who is living well and paying his taxes need not accept the extension agent's advice about new farming methods, but when his bottomland begins to play out, he is eager to try almost anything. The country that "wins" the war continues on its smug way of life, while the "loser" undergoes structural changes that cannot be accounted for in terms of "war damage." The student getting "straight A's" need not learn much, even if he is not being taxed to his capacity; whereas the student on "probation" is much more likely to consider carefully the path he has chosen for his life. Only when the "honor" student finds himself well into his profession will he find that dissatisfaction is pressing him toward changes that he is too deeply committed to make.

In biology, the overhead—the cost of regulation—is expressed in Selye's concept of "stress." The Used Car Law applied to biological situations says that the organism that keeps down the total stress need not adapt, but that the organism *will* adapt, or collapse, once the stress reaches an unacceptable level. The appearance of a change, therefore, is a heuristic signal to the observer that the cost of regulation has increased, even if the increase has not previously been noticed.

The same holds true in psychiatry, where the analysis often works backwards from adaptations to the stresses that underlie them. Such a heuristic is necessary because of the seemingly arbitrary nature of what might cause psychological stress. The Law of Effect does not seem to hold for the psychiatrist's systems. Some people are severely stressed by the sight of cigarettes in an ash tray, while others are sexually aroused by the sight of an untied shoelace. On the other hand, some people experience no emotion when frying people with napalm—something most of us would regard as more than sufficient cause for internal stress.

As an example of the difficulties facing a psychiatrist, consider the sense of self—or ego—that is often closely bound up with particular, or peculiar, behavior patterns. If a man identifies himself as "short-tempered," he may continue to throw tantrums even when signals from his environment seem (to us) to make such outbursts inappropriate. He may continue to behave in this socially unacceptable way even though it endangers his job, his friendships, or his family relationships. To him, the tantrums represent his identity—not the things whose loss is threatened. He filters the warning signals from his environment, so from our point of view, he is regulating to avoid adaptation, or change in behavior. From his point of view, however, he is regulating to preserve his identity, to survive. The more effective is this regulatory system, the less likely he is to change the offending behavior. The only hope for change is to change his method of identification, or substantially to increase his pain.

This example from psychopathology should not be lost upon us just because we feel so superior. Each of us clenches all sorts of beliefs about the world that we are equally reluctant to drop. Because they happen to be socially acceptable beliefs, we are not under obvious stress to release them. This lack of stress may be cause for optimism. If we have no great committment to "things we know that ain't so," we should be able to release a few of them without reclining on the psychiatrist's couch.

For instance, we ordinarily identify things in the world in such casual and intuitive ways that we are not prepared to discuss adaptation intelligently. We may mistake the physical shell for the system. The alumni association can convince the old grads that it's the *same* alma mater by sending them water colors of the old buildings—and painting the students in a vague way so as not to show the long hair. Students are just the ephemera of the university: its identity resides in the ivy-covered buildings and the ivy-covered professors.

Or worse yet, the ivy-covered *name*. What whopping changes slip under our noses so long as the name's the same. Political parties are the most notable examples, with "Bolshevik" and "Democratic" leading the parade. This book itself is a sterling example: though not a single page of the original manuscript has escaped the waste basket, the title has remained, and therefore it has remained the "same book."

As long as we are this arbitrary in our identification of systems, we cannot hope to understand much in general about their behavior. Yet we are perfectly entitled to identify systems in any way we choose. The recipe for effective thinking is to use those ways of identifying systems that focus on what interests us, and to discard those ways that do not. The "stress" that the Used Car Law reduces may simply be stress on the part of the observer, the mental cost of having a viewpoint too far out of touch with the "realities"—either of the world "out there" or of his own mind. We might, therefore, wish to rephrase the Used Car Law to put specific emphasis on the observer's role:

1. A way of looking at the world that is not putting excessive stress on an observer need not be changed.

2. A way of looking at the world may be changed to reduce the stress on an observer.

In other words, why do we continue pumping gas into certain antique ways of looking at the world, why do we sometimes expend mammoth efforts to repair them, and why do we sometimes trade them in? What better systems questions to add to our list as we conclude our introduction to general systems thinking?

QUESTIONS FOR FURTHER RESEARCH

1. *General*

Make a list of some systems that you encounter in your daily life, such as a particular tree, forest, animal, business, machine, building, river, or road. Try to establish a lifetime for each, and discuss the problems of attributing lifetimes to such a system.

Reference: Aldous Huxley, *After Many a Summer Dies the Swan.*
New York: Harper, 1939

in which there is an enormous amount of discussion attempting to figure out how long carp live in a pond.

2. *Computer Simulation*

In our OCCULT system, the transformation might be represented by a 10 by 10 table, showing at each intersection what digit was to be produced. The upper corner of the table, for example, would like this:

	0	1	2	3 ⋯
0	0	0	0	0 ⋯
1	0	1	2	3 ⋯
2	0	2	4	6 ⋯
3	0	3	6	9 ⋯
4	0	4	8	2 ⋯

But *any* such table would be suitable as a transformation, and, in particular, all or part of the digits of the *state* could overlap the table. Write a computer program to simulate this system, where the state and transformation table overlap, and study and report on its behavior.

3. *Political Science*

Comment upon the following statement about "equilibrium"[20]:

A society in perfect equilibrium might be defined as a society every member of which had at a given moment all that he could possibly desire and was in a state of absolute contentment; or it might be defined as a society like that of certain social insects such as bees or ants, in which every member responds predictably to a given stimuli. Obviously any human society can be in but an imperfect equilibrium, a condition in which the varying and conflicting desires and habits of individuals and groups are in complex mutual adjustment, an adjustment so complex that no mathematical treatment of it seems possible at present.

4. *Role Theory*

In the identification of human beings, we often employ the strategy of "roles." If we ask, "who is that?" we *may* get the answer, "Harley Krank." But we are more likely to hear "the mailman," or ".the butcher's assistant," or "the next-door neighbor." Role theory is a highly developed branch of sociology or social psychology, and can be related in numerous ways to the questions of identity raised in this chapter. Discuss the relationship between identifying variables, identifying behavior, and the "role" of role theory.

Reference: David Krech, Richard Crutchfield, and Egerton L. Ballachey, *Individual in Society.* New York: McGraw-Hill, 1962.

5. *Welfare*

In 1704, Daniel Defoe published a pamphlet whose content was well described by its title:

"Giving Alms no Charity and employing the Poor a Grievance to the Nation"

Many times subsequently, the same argument has been advanced: that if the income of the workers is *regulated,* then they need not *adapt* by accepting lower wages or harder work. By taking away the regulatory mechanisms—poor laws, and the like—the worker could be made to choose between adapting or failure to survive—starving, to put it plainly. Discuss the assumptions underlying this argument, in terms of general systems ideas.

6. *Diplomacy*

Although the idea existed certainly as far back as the Greeks, the first modern statement of the idea of "Balance of Power" was given by Hume in 1854. Actually, however, a balance-of-power system had been operating at that time for over two hundred years in Europe, effectively since the Treaty of 1648. Perhaps it is significant that only when the system began to show signs of collapse did we begin to get explicit recognition of its existence by political theorists.

Discuss the balance of power as a regulatory scheme. What events would indicate an adaptation of such a system? What would it mean for the balance of power to fail to survive? Does the system still exist in Europe today?

Reference: Karl Polanyi, *The Great Transformation.* Boston, Mass.: Beacon Press, 1964.

7. *Picture Processing*

Make a list of common normalizing transformations we apply to pictures, such as rotation, change of size, change of color, eliminating "smudges," straightening lines, or closing gaps. Discuss various normalizations that can be composed of combinations of these transformations giving examples of pictures each normalization will classify as "same" and "different." Give examples in which the sequence in which the normalizations are applied makes a difference in the definition of sameness.

8. *Language*

Answer the following question:

One of the most obvious characteristics of any concrete individual word is that it is able to appear in any style, in any color, in any size, in shorthand or longhand. If it is spoken, it may be uttered with any audible intensity, fast or slow, at any pitch, and, in most languages, with many intonations. All these so various manifestations may be of the "same" word. In what, now, does this sameness consist?[21]

9. *Neurology and Education*

In recent years, there has been increasing controversy over the practice of giving certain drugs—particularly amphetamines—to schoolchildren, diagnosed as having a "disorder" known as "minimal brain dysfunction—MBD." The diagnosis of this syndrome is done strictly by behavior: earlier, the syndrome was attributed to "brain damage," but nobody succeeded in discovering any structural "damage" to the brain. The problem in diagnosis comes from the broad range of symptoms, all of which are "normal" in some circumstances, which may be used as symptoms of this syndrome. Such symptoms are: poor academic achievement, impulsive acts, short attention span, sudden emotional changes, lack of coordination, hyperactivity, and ease of distraction. It can be seen from this list how easy it would be to assign the label "MBD" to any student who does not conform to the teacher's idea of what a good student should be like. Discuss the problems of such a behavioral identification of "disease," particularly in the case of MBD.

> *Reference:* Paul H. Wendler, *Minimal Brain Dysfunction in Children.* New York: Wiley, 1971.

10. *Ecology*

Discuss the following remark:

A significant difference that can be found between a steam engine and the mouse population or any other appropriate biological working machine is that the living system must use part of its enery to manufacture and repair itself.

> *Reference:* Lawrence B. Slobodkin, *Growth and Regulation of Animal Populations,* p. 132. New York: Holt, Rinehart, and Winston, 1966.

11. *Medicine*

Discuss the following remarks:

Most discussions of death and dying shift uneasily, and often more or less unconsciously, from one point of view to another. On the one hand, the common noun "death" is thought of as standing for a clearly defined event, a step function that puts a sharp end to life. On the other, dying is seen as a long-drawn-out process that begins when life itself begins and is not completed in any given organism until the last cell ceases to convert energy.[22]

See also:

> Leon R. Kass, "Death as an Event: A Commentary on Robert Morison." *Science,* **173,** 698 (August 20, 1971).

12. *Oceanography or Marine Biology*

Sharks have been around for 200 million years, although they have shrunk considerably from their former length of 60 to 70 feet. Their remarkable abilities to adapt to their environment put every creature but the cockroach out of their league.

How is it that sharks are still sharks though they have shrunk by perhaps a factor of 10 in length? Would people still be people if in 200 million years they are as tall in inches as they now are in feet?

Constance Holden, "Shark-Tagging: Keeping Track of One of the World's Great Survivors. *Science,* **180** (25 August 1972).

13. *The Ultimate Question*

Will people still be around in 200 million years?

READINGS

RECOMMENDED

1. R. W. Gerard, "Units and Concepts of Biology." *Modern Systems Research for the Behavioral Scientist,* Walter Buckley, Ed., pp. 51–58. Chicago: Aldine, 1968.
2. Robert S. Morison, "Death: Process or Event?" *Science,* **173,** 694 (August 20, 1971).

   Leon R. Kass, "Death as an Event: A Commentary on Robert Morison." *Science,* **173,** 698 (August 20, 1971).

SUGGESTED

1. Hans Selye, *The Stress of Life.* New York: McGraw-Hill, 1956.
2. Mikhael Bakunin, *God and the State.* New York: Dover, 1970.

Notes

CHAPTER 1

1. John R. Platt, "Strong Inference." *Science,* **146,** No. 3642, 351 (1964).
2. Paul W. Richards, "The Tropical Rain Forest." *Scientific American,* Vol. 229, #6, Dec. 1973, pp. 58–67.
3. Eugene P. Wigner, Nobel Prize Acceptance Speech, December 10, 1963. Reprinted in *Science,* **145,** No. 3636, 995 (1964).
4. Karl Deutsch, "Mechanism, Organism, and Society." *Philosophy of Science,* **18,** 230 (1951).
5. Anatol Rapoport, "Mathematical Aspects of General Systems Analysis." *General Systems Yearbook,* **XI** 3 (1966).
6. W. Ross Ashby, "Systems and Their Information Measures." *Trends in General Systems Theory,* George J. Klir, Ed., pp. 78–97 New York: Wiley, 1971.
7. Richard Feynman, *The Character of Physical Law.* Cambridge, Massachusetts: MIT Press, 1965.

   Feynman actually quotes this remark without attribution, but the real reason for referring to Feynman is to contrast the physicist's view of this discovery with the point of view taken here. Feynman says:

   . . . I am interested not so much in the human mind as in the marvel of a nature which can obey such an elegant and simple law as this law of gravitation. Therefore our main concentration will not be on how clever we are to have found it all out, but on how clever nature is to pay attention to it.   (p. 14):

   Our interest, of course, is precisely the opposite, which means that Feynman is a good supplement to this chapter and the next.

8. Erwin Schrödinger, *What is Life?* Cambridge: Cambridge University Press, 1945.
9. Ludwig von Bertalanffy, *General Systems Theory,* p. 49. New York: Braziller, 1969. Copyright © 1969 by George Braziller, Inc. Reprinted with the permission of the publisher.
10. D'Arcy Thompson, *On Growth and Form,* abridged ed., John Taylor Bonner, Ed., pp. 262–263. Cambridge: Cambridge University Press, 1961.

CHAPTER 2

1. Robinson Jeffers, "The Answer." In *The Selected Poetry of Robinson Jeffers.* New York: Random House, 1937.
2. Geraldine Colville and H. M. Colville, *Matisse, From the Life,* p. 124. London: Faber and Faber, 1960.

3. J. W. S. Pringle, "On the Parallel Between Learning and Evolution." In *Modern Systems Research for the Behavioral Scientist,* Walter Buckley, ed., pp. 259–280. Chicago: Aldine, 1968.

4. Claude Bernard, *An Introduction to the Study of Experimental Medicine.* New York: Dover, 1957.

5. Edward T. Hall, *The Silent Language.* Garden City, N.Y.: Doubleday, 1959.

6. Thomas S. Kuhn, *The Structure of Scientific Revolutions.* Chicago: University of Chicago Press, 1962.

7. Max Planck, *Scientific Autobiography, and Other Papers.* New York: Philosophical Library, 1949.

8. Hans Reichenbach, *The Rise of Scientific Philosophy.* Berkeley, Calif.: University of California Press, 1963.

9. Hans Selye, *The Stress of Life.* New York: McGraw-Hill, 1956.

10. Kenneth Boulding, "General Systems as a Point of View." In *Views on General Systems Theory,* Mihajlo D. Mesarovic, Ed. New York: Wiley, 1964.

11. Hans Reichenbach, *Op. cit.*

12. Kenneth Boulding, "General Systems as a Point of View," *op. cit.*

13. Jean Piaget, *The Language and Thought of the Child.* Cleveland: World, 1955.

14. Selye, *Op. cit.*

15. Max Wertheimer, Ed., *Productive Thinking,* pp. 269–70. New York: Harper, 1959. Copyright 1945, 1959 by Valentin Wertheimer.

16. Karl Menninger, *Theory of Psychoanalytic Technique,* p. 14. New York: Basic Books, 1958.

17. Anatol Rapoport, In *Modern Systems Research for the Behavioral Scientist,* Walter Buckley, Ed., p. xiii. Chicago: Aldine, 1968.

18. Alfred North Whitehead, *Science and the Modern World.* New York: Macmillan, 1926.

19. Kenneth Boulding, "General Systems as a Point of View," *op. cit.*

20. Mark Kac, "Some Mathematical Models in Science." *Science,* **166,** No. 3906 695 (1969).

21. Paul Samuelson, *Economics* (eighth edition), pp. 19–23. New York: McGraw-Hill, 1970.

22. Gerald M. Weinberg, "Systems Research Potentials Using Digital Computers." *General Systems Yearbook,* **VIII,** 145 (1963).

23. Gerald M. Weinberg, "Natural Selection as Applied to Computers and Programs." *General Systems Yearbook,* **XV,** 145 (1970).

24. Daniela Weinberg, "Models of Southern Kwakiutl Social Organization." In *Cultural Ecology and Canadian Native Peoples,* Bruce Cox, Ed. Carleton Library Series, Institute of Canadian Studies. Ottawa: Carleton

University. (1974)

25. G. M. Weinberg and Daniela Weinberg, "Biological and Cultural Models of Inheritance." *General Systems Journal,* **I,** No. 2 (1974).

26. Gerald M. Weinberg, *The Psychology of Computer Programming.* New York: Van Nostrand Reinhold, 1971.

27. Donald Gause and G. M. Weinberg, "On General Systems Education." *General Systems Yearbook,* **XVIII,** 137 (1973).

28. Ludwig von Bertalanffy and Anatol Rapoport Ed., *General Systems Yearbook.* Vols. 1–19. Ann Arbor: Society for General Systems Research, 1956–1974.

29. Ludwig von Bertalanffy *General Systems Theory.* New York: Copyright © 1968 by George Braziller, Inc. Reprinted with the permission of the publisher.

CHAPTER 3

1. W. Ross Ashby, "Principles of the Self-Organizing System." In *Modern Systems Research for the Behavioral Scientist,* Walter Buckley, Ed. Chicago: Aldine, 1968.

2. Robert Herrick, "Delight in Disorder." In *The Complete Poetry of Robert Herrick.* Garden City, N.Y.: Doubleday, 1963.

3. Simone de Beauvoir, *Memoirs of a Dutiful Daughter.* Baltimore: Penguin Books, 1958.

4. Albert Einstein, "Maxwell's Influence on the Evolution of the Idea of Physical Reality." 1931 (this was the first sentence of the essay).

5. E. H. Gombrich, *Art and Illusion,* no. 5 in the A. W. Mellon Lectures in the Fine Arts, Bollinger Series XXV © 1960, 1961, and 1969 by The Trustees of the National Gallery of Art, Washington, D.C., reprinted by permission of Princeton University Press.

6. Eleanor Gibson, "The Development of Perception as an Adaptive Process." *American Scientist,* **58,** 98 (1970).

7. Ward H. Goodenough, *Culture, Language, and Society.* Reading, Mass.: Addison-Wesley, 1971.

8. James G. Miller, "Living Systems: The Organization." *Behavioral Science,* **17,** No. 1, 19 (1972).

9. Clarence Lewis and Cooper Langford, *Symbolic Logic,* p. 256 Peter Smith, 1959.

10. L. A. Zadeh, "Fuzzy Sets." *Information and Control,* **8,** 338 (1965).

11. Arthur D. Hall and R. E. Fagen, "Definition of System." In *Modern Systems Research for the Behavioral Scientist,* Walter Buckley, Ed. Chicago: Aldine, 1968.

12. We shall introduce whatever set concepts we need, but the reader may wish to boneup on his set theory separately. Some good references are:

P. R. Halmos, *Naive Set Theory.* Princeton, N.J.: Van Nostrand, 1960.

S. Lipschutz, *Set Theory and Related Topics.* New York: Schaum, 1964.

Zadeh, *Op. cit.*

W. Ross Ashby, *An Introduction to Cybernetics.* New York: Wiley, 1961.

W. Ross Ashby, "The Set Theory of Mechanism and Homeostasis." *General Systems,* **IX,** 83 (1964).

13. Antoine de Saint-Exupéry, *Le Petit Prince,* p. 19. Paris: Gallimard, (Author's translation).

14. S. S. Stevens, "Mathematics, Measurement, and Psychophysics." In *Handbook of Experimental Psychology,* S. S. Stevens, Ed. New York: Wiley, 1962.

15. Crane Brinton, *The Anatomy of Revolution,* pp. 178–9. New York: Vintage Press, 1965.

CHAPTER 4

1. W. Ross Ashby, *Introduction to Cybernetics.* New York: Wiley, 1961.

2. G. M. Weinberg, "Learning and Meta-Learning Using a Black Box." *Cybernetica,* **XIV,** No. 2 (1971).

3. Morton H. Fried, *The Study of Anthropology.* New York: Crowell, 1972.

4. John R. Dixon and Alden H. Emery, Jr., "Semantics, Operationalism, and the Molecular-Statistical Model in Thermodynamics." *American Scientist,* **53,** 428 (1965). Reprinted by permission of *American Scientist* journal of Sigma Xi, The Scientific Research Society of North America.

5. R. E. Gibson, "Our Heritage from Galileo Galilei." *Science,* **145,** 1271 (September 18, 1964).

6. Robert R. Newton, *Ancient Astronomical Observations and the Accelerations of the Earth and Moon.* Baltimore: Johns Hopkins Press, 1970.

7. Kenneth Boulding, *Economics as Science,* p. 115. New York: McGraw-Hill, 1970. Used with permission of McGraw-Hill Book Company.

8. James C. Maxwell, Source lost and untraceable.

9. Neils Bohr, *Essays, 1958–1962.* New York: Wiley, 1963.

10. Bohr, *Op. cit.*

11. Robert Redfield, *Tepotzlan: A Mexican Village.* Chicago: University of Chicago Press, 1930.

12. Oscar Lewis, *Life in a Mexican Village: Tepotzlan Restudied.* Urbana: University of Illinois Press, 1951.

13. Oskar Morgenstern, *On the Accuracy of Economic Observations.* New Jersey: Princeton University Press, 1963.

14. Thomas R. Blackburn, "Sensuous–Intellectual Complementarity in Science." *Science,* **172,** 1003 (June 4, 1971).

15. James Loy, Review of *Social Groups of Monkeys, Apes, and Men,* by M. R. A. Chance and C. J. Jolly. *Science,* **172** (May 1971).

16. W. M. Elsasser, "Quanta and the Concept of Organismic Law." *Journal of Theoretical Biology,* **1,** 27 (1961).

CHAPTER 5

1. Kurt Vonnegut, Jr. *Cat's Cradle,* pp. 67–68. New York: Dell, 1970. Copyright © 1963 by Kurt Vonnegut, Jr. Reprinted by permission of the publisher, Delacorte Press/Seymour Lawrence.

2. Henry P. Bowie, *On the Laws of Japanese Painting.* Gloucester, Mass.: Peter Smith, 1911.

3. Leo Tolstoy, *Childhood, Boyhood, and Youth.* New York: McGraw-Hill, 1965. Used with permision of McGraw-Hill Book Company.

4. Elliott Jaques, *The Changing Culture of a Factory.* London: Tavistock, 1951.

5. Galileo, "Dialogo," *Opere,* **VII,** p. 129. In Herman Weyl, *Philosophy of Mathematics and Natural Science,* p. 16. New York: Atheneum, 1963.

6. Hermann Hesse, *Magister Ludi,* In *Eight Great Novels of H. Hesse.* New York: Bantam Press, 1972.

7. In *One Hundred Poems from the Japanese,* p. 51, Kenneth Rexroth, Ed. and Trans. New York: New Directions, 1959.

8. Oskar Morgenstern, *On the Accuracy of Economic Observations.* New Jersey: Princeton University Press, 1963.

9. P. W. Bridgman, *The Way Things Are,* p. 109. Cambridge, Mass.: Harvard University Press, 1959.

10. Ernst Mayr, *Animal Species and Evolution.* Cambridge, Mass.: Harvard University Press, 1963.

CHAPTER 6

1. P. W. Bridgman, *The Way Things Are,* p. 3. Cambridge, Mass.: Harvard University Press, 1959.

2. Herbert A. Simon, *The Sciences of the Artificial,* p. 18. Cambridge, Mass.: MIT Press, 1969.

3. Henry L. Langhaar, *Dimensional Analysis and Theory of Models.* New York: Wiley, 1951.

4. A good place to start the study of analog computers is: J. R. Ashley, *Introduction to Analog Computation.* New York: Wiley, 1963.

5. G. M. Weinberg, N. Yasukawa, and R. Marcus, *Structured Programming in PL/C.* New York: Wiley, 1973.

6. W. Ross Ashby, *Introduction to Cybernetics.* New York: Wiley, 1961.

7. The subject of topology is not ordinarily presented in a way that makes it accessible to the mathematically unwashed. The interested reader may wish to examine the discussion by Richard Courant & Herbert Robbins in *Topology: The World of Mathematics,* James R. Newman, Ed.

New York: Simon and Schuster, 1956–1960 (4 volumes). Those with more mathematics but no topology might try: M. Mansfield, *Introduction to Topology.* Princeton, N.J.: Van Nostrand, 1963.

8. Hans Elias, "Three-Dimensional Structure Identified from Single Sections." *Science,* **174** 993 (December 3, 1973).
W. A. Gaunt, *Microreconstruction.* London: Pitman Medical Press, 1971.

9. Edwin Abbott, *Flatland: A Romance in Many Dimensions.* New York: B & N Press, 1963.

10. George Kirkland, farm worker, as quoted by Ronald Blythe, *Akenfield: Portrait of an English Village,* p. 99. Middlesex, England: Penguin Books, 1972.

11. This and innumerable other interesting observations on the nature of time can be found in Leonard W. Doob, *Patterning of Time.* New Haven: Yale University Press, 1972.

12. See, for example, Murray R. Spiegel, *Laplace Transforms.* New York: Schaum, 1965.

13. Ingrid U. Olsson, Ed., "Nobel Symposium 12: Radio-Carbon Variations and Absolute Chronology." New York: Wiley, 1970.

14. Notice the variation in which two variables are plotted against time on the same graph—a clever way to ensure equal time scales.

15. L. Brillouin, "Life, Thermodynamics, and Cybernetics." *Modern Systems Research for the Behavioral Scientist,* p. 149, Walter M. Buckley, Ed. Chicago: Aldine, 1968.

16. Nikos Kazantzakis, *The Last Temptation of Christ.* New York: Simon and Schuster, 1966.

17. Norman Howard-Jones, "The Origins of Hypodermic Medication." *Scientific American,* (January 1971).

18. J. Woodland Hastings, "Light to Hide by: Ventral Luminescence to Camouflage the Silhouette." *Science,* **173,** 116 (Sept. 10, 1971).

19. Carl F. Jordan, "A World Pattern in Plant Energetics." *American Scientist,* **59,** 425 (July–August 1971).

CHAPTER 7

1. R. W. Gerard, "Units and Concepts in Biology." *Modern Systems Research for the Behavioral Scientist,* Walter Buckley, Ed., pp. 51–58. Chicago: Aldine, 1968.

2. Mikhael Bakunin, *God and the State.* New York: Dover, 1970.

3. R. F. Dabenmire, *Plants and Environment,* p. 272. New York: Wiley, 1959.

4. Charles Gregg, Ed., *American Environmental Studies.* (Forty-two volumes) New York: Arno Press, 1970.

5. L. A. Zadeh and C. A. Desoer, *Linear System Theory.* New York: McGraw-Hill, 1963. Stability in nonlinear systems of various types is, of course, a more difficult topic mathematically. The interested reader may find the following reference helpful. (Prerequisites: matrices, differential equations, and linear systems.)
Jack M. Holtzman, *A Functional Analysis Approach.* Englewood Cliffs, N.J.: Prentice-Hall, 1970.

6. W. J. Cunningham, "The Concept of Stability." *American Scientist,* **53,** 431 (December 1963). Reprinted by permission of *American Scientist* journal of Sigma Xi, The Scientific Research Society of North America.

7. T. Parsons and E. A. Shils, *Toward a General Theory of Action,* p. 107. Cambridge, Mass.: Harvard University Press, 1951.

8. Michael A. Duguay, "Light Photographed in Flight." *American Scientist,* **59,** 550 (Sept.–Oct. 1971).

9. Charles Darwin, *On the Origin of Species* (Facsimile edition). Cambridge, Mass.: Harvard University Press, 1964.

10. Julian Steward, *Theory of Culture Change,* p. 184. Urbana: University of Illinois Press, 1963.

11. John Von Neumann, *Theory of Self-Reproducing Automata.* Urbana: University of Illinois Press, 1966.

12. W. Ross Ashby, *Introduction to Cybernetics,* p. 9. New York: Wiley, 1961.

13. See, for example, Harry C. Andres, *Introduction to Mathematical Techniques in Pattern Recognition.* New York: Wiley, 1972.
Paul A. Kolers and Murray Eden, Eds., *Recognizing Patterns: Studies in Living and Automatic Systems.* Cambridge: MIT Press, 1968.

14. See, for example: R. Duda and P. Hart, *Pattern Classification and Scene Analysis.* New York: Wiley, 1973.

15. William Butler Yeats, "Among School Children." From *Collected Poems.* New York: Macmillan, 1956.

16. D. O. Hebb, *The Organization of Behavior.* New York: Wiley, 1949.

17. Herbert Spencer, *The Principles of Sociology,* pp. 447–448. New York: Appleton-Century-Crofts, 1904.

18. Nils J. Nilsson, *Learning Machines.* New York: McGraw-Hill, 1965.

19. Hans Selye, *The Stress of Life,* p. 54. New York: McGraw-Hill, 1956. Used with permission of McGraw-Hill Book Company.

20. Crane Brinton, *The Anatomy of Revolution,* pp. 15–16. New York: Vintage Books, 1965.

21. P. W. Bridgman, *The Way Things Are,* p. 13. Cambridge: Harvard University Press, 1959.

22. Robert S. Morison, "Death: Process or Event?" *Science,* **173,** 694 (August 20, 1971).

Appendix

Name of Notation	Symbols	Read as
A. Scientific notation	10^{15}	1. Ten to the fifteenth 2. One with fifteen zeros 3. 1,000,000,000,000,000
	6×10^{10}	1. Six times ten to the tenth 2. Six with ten zeros 3. 60,000,000,000
	10^{-15}	1. Ten to the *minus* fifteenth 2. One with *fourteen* $(15 - 1)$ zeros *in front of it.* 3. .000 000 000 000 001 (In other words, a *small* number.)
B. Subscript notation	a_2	1. *a*-sub-two 2. The second element of a
	P_i	1. *P*-sub-*i* 2. The "*i*th" (pronounced "eyeth") element of P.
(two-dimensions)	M_{32}	1. *M*-sub-three-two 2. The element in *row* 3, *column* 2, of the matrix M.
	X_{ij}	1. *X*-sub-*i*-*j* 2. The element in the *i*th row and *j*th column of X.
C. Sets	(A, X, H)	1. The set of elements named "*A*," "*X*," and "*H*." 2. The set of elements A, X, and H. 3. The set A, X, H.
	$(A, B, C, \ldots)$	The set consisting of A, B, C, *and so forth.*
	$((a, v), (a, b),$ $(b, y))$	1. The set of sets, (a, v), (a, b), (b, y). 2. The set of *pairs*, (a, v), (a, b), (b, y).
	$S = (r, t, q, c)$	1. S is the name of the set containing r, t, q, and c. 2. S is the set (r, t, q, c)
D. Cartesian product	$S \times T$	1. The Cartesian product of set S and set T. 2. S cross T. 3. The set of all possible ordered pairs composed of one element from the set S followed by one from T. 4. The product set of S and T.

E. Subset or partition $S = (a,b,(r,t))$

1. S is the set consisting of the elements a and b and the subset (r, t).
2. S is *partitioned* into a, b, and (r,t).

F. Functional notation $y = f(a,d)$

1. y depends (only) on a and d, but in some unspecified way.
2. y depends on a and d
3. y is a function of a and d.
4. y equals f-of a and d.

$r = g(e,t, \ldots)$

r is a function of e, t, and possibly some other variables.

$s = g(x) + h(x,z)$

s is some function of x alone added to some function of x and z.

$y = f(a, d)$
$y = F(a, d)$
$y = \phi(a, d)$
$y = \theta(a, d)$
$y = \psi(a, d)$

All these have the same meaning: y depends (only) on a and d, but in some unspecified way.

Note: The most common symbols for "some unspecified function" are the small letters f, g, h; the capital letters F, G, H; and the Greek letters, ϕ(phi), θ(theta), and ψ(psi). Nevertheless, since the symbol in that position stands for "some unspecified function," any symbol might be used.

Author Index

Subject Index